The Handling of Chromosomes

Sixth Edition

Other works by Professor C. D. Darlington

CHROMOSOMES AND PLANT BREEDING
Macmillan, 1932

RECENT ADVANCES IN CYTOLOGY
Churchill, 1932

THE EVOLUTION OF GENETIC SYSTEMS
Cambridge University Press, 1939
Oliver and Boyd, 1958

THE FACTS OF LIFE
George Allen & Unwin, 1953

CHROMOSOME BOTANY AND THE ORIGINS OF CULTIVATED PLANTS
George Allen & Unwin, 1963

GENETICS AND MAN
George Allen & Unwin, 1964

CYTOLOGY (3RD ED.)
Churchill, 1965

THE CONFLICT OF SCIENCE AND SOCIETY
Watts, 1948

DARWIN'S PLACE IN HISTORY
Blackwell, 1959

THE EVOLUTION OF MAN AND SOCIETY
George Allen & Unwin, 1969

with K. Mather
THE ELEMENTS OF GENETICS
George Allen & Unwin, 1961

GENES, PLANTS AND PEOPLE
George Allen & Unwin, 1959

with E. K. Janaki Ammal
CHROMOSOME ATLAS OF CULTIVATED PLANTS
George Allen & Unwin, 1945

with A. P. Wylie
CHROMOSOME ATLAS OF FLOWERING PLANTS
George Allen & Unwin, 1956

with A. D. Bradshaw (edited)
TEACHING GENETICS
Oliver and Boyd, 1963, 1966

The Handling of Chromosomes

Sixth Edition

C. D. DARLINGTON, F.R.S.
and
L. F. LA COUR, F.R.S.

Revised by L. F. La Cour

London   George Allen & Unwin Ltd
Ruskin House   Museum Street

First published in 1942
Revised Second Edition 1947
Second Impression 1950
Third Revised Edition 1960
Fourth Revised Edition 1962
Second Impression 1966
Fifth Edition 1969
Second Impression 1970
Sixth Edition 1976

ISBN 0 04 574014 3

Printed in Great Britain
in 10 point Times Roman type
by Clarke, Doble & Brendon Ltd
Plymouth

To
JOHN BELLING
1866–1933
whose ingenious invention
brought the chromosomes within
reach of every enquirer

PREFACE TO THE SIXTH EDITION

The chromosomes are responsible for heredity and variation and they control development. Their study has therefore become necessary for teaching and research in all parts of the life sciences. This book attempts to show how they have to be handled for such purposes.

The methods range from the very simplest treatments to the most elaborate. The present edition has been fully revised to cover the advances of the last six years, particularly in the study of mammalian chromosomes, as well as the means to reveal heterochromatin whether by Giemsa banding or quinacrine fluorescence.

It brings up to date all schedules, calendars and classifications included in the previous edition. There are two new plates and the Bibliography has been extensively enlarged.

C. D. Darlington   L. F. La Cour
Botany School   *School of Biological Sciences University of*
Oxford   *East Anglia Norwich Norfolk*

ACKNOWLEDGEMENTS

We wish to thank the following collaborators for data, methods, references and photographs:

Professor P. C. Koller   Chester Beatty Institute, London S.W.3

Professor H. G. Callan, FRS   Zoology Department, University of St Andrews

Professor P. T. Thomas   Director of the Welsh Plant Breeding Station, Aberystwyth

Professor D. Lewis, FRS   Botany Department, University College, London

Professor Bernard John   Research School of Biological Sciences, Australian National University, Canberra, ACT, Australia

Professor D. G. Catcheside, FRS   John Curtin School of Medical Research, University of Canberra, ACT, Australia

Mr G. E. Marks   John Innes Institute, Colney Lane, Norwich

Dr C. E. Ford, FRS   Member, External Staff, Medical Research Council, Sir William Dunn School of Pathology, Oxford

Professor H. John Evans   MRC Clinical and Population Cytogenetics Unit, Western General Hospital, Edinburgh

Dr Marco Fraccaro   Population Genetics Unit, Headington

Dr D. Davidson, Dr K. R. Lewis and Dr C. G. Vosa   Botany School, Oxford

For the preparation of the photographs and diagrams continued from previous editions, we are indebted to our late colleague and friend Mr H. C. Osterstock.

C.D.D.   L.F.La C.

CONTENTS

Preface   *page*   9

1 ORIGIN, SCOPE AND PURPOSE OF CHROMOSOME WORK   19

2 EQUIPMENT   22
   a. Microscopes   22
   b. Resolving Power and Magnification   23
   c. Lighting and Screens   25
   d. Oil and Lenses   26
   e. Camera Lucida   27
   f. Other Equipment   27

3 LIVING CHROMOSOMES   28
   a. Material   28
   b. Methods   28
   c. Vital Staining   29
   d. Chromosome Optics   29

4 BULK FIXATION   31
   a. Principles   31
   b. Practice   32
   c. Mordanting and Hydrolysis   33
   d. Penetration   34
   e. Condition of the Material   34
   f. Duration   35
   g. Uses of Bulk Fixation   35

5 SMEARS AND SQUASHES   36
   a. Material   36
   b. Smear Methods   36
   c. Acetic Squash Methods   36
   d. Maceration   39

6 PARAFFIN METHODS   41
   a. Value   41
   b. Fixation and Washing   41
   c. Dehydration and Infiltration   41
   d. Embedding   42
   e. Section Cutting   43
   f. Mounting Ribbons   43
   g. Removing the Wax   44

7 STAINING AND MOUNTING   45
   a. Staining   45
   b. Mounting   48
   c. Fading   48
   d. Re-staining   49
   e. Bleaching   49

12 Contents

8 SPECIAL TREATMENTS   *page* 50
   a. Pre-Treatment for Structure   50
   b. Micro-Chemical Tests   51
   c. Animal Eggs   55
   d. Salivary Glands   56
   e. Tadpole Tails   57
   f. Amphibian Testes   58
   g. Lampbrush Chromosomes   58
   h. Culex Testes   59
   i. Blood   59
   j. Mammalian Embryos   60
   k. Tumours   60
   l. Diagnosis of Sex in Somatic Tissue   61
   m. Giemsa Banding in Metaphase Chromosomes   61
   n. Recognition of Heterochromatin with Fluorochromes   63
   o. Identification of Sister Chromatids by Differential Staining 63
   p. In Situ Localisation and Characterisation of Different Classes
      of Chromosomal DNA   64
   q. Embryo Sac and Endosperm   65
   r. Pre-Meiotic Mitosis   65
   s. Pachytene   65

9 THE CONTROL OF MITOSIS   67
   a. Irradiation   67
   b. Chemical Agencies   76
   c. Differential Reactivity   79
   d. Temperature   80
   e. Centrifuging   81
   f. Various Genetical Devices   81

10 THE CONTROL OF FERTILISATION   82
   a. Pollen Germination   82
   b. Tube Division   83
   c. Pollen Storage   83
   d. The Style   84
   e. Haploid Plants   85
   f. Haploid Animals   86

11 PHOTOGRAPHY   87
   a. Uses   87
   b. Cameras   87
   c. Taking the Photograph   88
   d. Developing   88
   e. Printing and Regulation of Contrast   89
   f. Preparing Photographs for Reproduction   90
   g. Screen Projection   90

12 AUTORADIOGRAPHY   91
   a. Uses   91
   b. Methods   91
   c. Preparation of Tissues   92
   d. Exposure of Autoradiographs   93
   e. The Amount of Tracer Necessary to
      Obtain Autoradiographs   93
   f. Background and Artefacts   94

g. *Observation of Autoradiographs*   *page* 95
h. *Removing an Autoradiograph*   95

13 DESCRIBING THE RESULTS   98
a. *Interpretation*   98
b. *Illustration*   99
c. *Description*   99

Appendix
I SOURCES OF MATERIAL   101
a. *Natural Groups*   102
b. *Chromosome Calendar*   107

II STANDARD SOLUTIONS   111
a. *Fixing Solutions*   111
b. *Stains for Chromosomes*   115
c. *Stains for Styles*   117
d. *Various Fluids*   117
e. *Drosophila Medium*   119
f. *Photographic Solutions*   120
g. *Transfer Solution for Autoradiography Stock Solution*   120

III SCHEDULES OF TREATMENT   121
1. *Paraffin Preparation*   121
1A. *Ester Wax Preparation*   122
2. *Smear Methods*   123
3. *Acetic-Lacmoid Squash Method*   124
3A. *Rapid Tumour Squashes*   124
4. *Acetic-Orcein Squash Method*   124
5. *Feulgen Squash Method*   125
5A. *Feulgen Squash Method for Endosperm*   126
6. *Feulgen Method for Sections and Smears*   126
6A. *Feulgen-Light-Green Method*   127
7. *Rapid Toluidine-Blue Squash Method*   127
8. *Quick-Freeze Method for Making Squash
Preparations Permanent*   128
9. *Crystal Violet Method*   128
10. *Rapid Haematoxylin Method*   129
11. *Giemsa-Gelei Method*   129
12. *Orange G-Aniline Blue Method*   129
13. *Azure B Method*   129
14. *Methyl-Green-Pyronin Method*   130
15. *Method for Using D N-ase*   130
16. *Modified Sakaguchi Reaction for Arginine*   131
17. *Buffered Thionin Method*   131
18. *Giemsa Banding Technique for Recognition of
Heterochromatin*   131
18A. *The Use of Trypsin for Mapping G-Bands*   132
19. *Quinacrine Fluorescence Method for Recognition of
Heterochromatin*   133
20. *Carbol Fuchsin Method*   133
21. *Tween Method for Nucleolar Structure*   134
22. *Autoradiography: Stripping-Film
Technique*   134

14 *Contents*

23. *Floating Cellophane Method for Pollen*   *page* 136
24. *Blood Culture Method for Human Chromosomes in
      Leucocytes from Whole Blood*   136
25. *Air-Dried Chromosome Preparations from Bone Marrow*   137
26. *Meiotic Preparations from Mammalian Testes*   138
27. *Identification of Sister Chromatids by Differential Staining* 139

IV CATALOGUE OF IMPLEMENTS   141

V   ABBREVIATIONS   145

VI GLOSSARY   146

    Bibliography   153
    A. Books and Reviews
    B. Special Articles

    Index   189

PLATES
after page 144

I   Pollen grains of *Paris*
II   Chromosome complements of animals
III   The birth of a B chromosome: *Agropyron*
IV   Nucleic acid starvation: *Trillium*
V   Colchicine mitosis: *Allium*
VI   Mitosis in blood precursor cells: Man
VII   Spontaneous breakage of chromosomes
VIII   Living salivary gland nucleus: *Chironomus*
IX   Polytene chromosome structure: *Drosophila*
X   Cleavage mitosis in a fish: *Coregonus*
XI   Embryo sac meiosis in a lily: *Fritillaria*
XII   Pachytene with nucleoli and heterochromatin
XIII   Diplotene chiasmata in *Fritillaria*
XIV   Metaphase chiasmata, free and localised: *Mecostethus* and *Fritillaria*
XV   Terminal localisation: *Chrysochraon* and *Triton*
XVI   Spiral structure at meiosis: *Tradescantia*
XVII   Ring formation: *Tradescantia* and *Rhoeo*
XVIII   Meiosis in a locust: *Pyrgomorpha*
XIX   Tetraploidy and interchange hybridity in a locust and a grasshopper
XX   Suppressed meiosis in a species cross: *Narcissus*
XXI   Ring-of-six in a cockroach: *Periplaneta*
    Diffuse centromeres in a rush: *Luzula*
XXII H^3 Autoradiograph of *Scilla sibirica*
XXIII   Sister chromatid exchange in chromosomes of Chinese hamster
XXIV   Lampbrush chromosomes in *Triturus*
XXV   G-banding in somatic chromosomes of plants:
    (i) *Anemone blanda*, (ii) *Fritillaria lanceolata*
XXVI   Meiosis in the bed bug: *Cimex*
XXVII   Meiosis in man: (i) Pachytene in a human spermatocyte
    (ii) first metaphase in the same individual
XXVIII   Tomato pachytene
XXIX   (i) Meiosis in a hybrid: *Tulbaghia violacae* and *T. dregeana*
    (ii) Nucleolar organisers in triploid endosperm: *Scilla sibirica*
XXX   Q- and G-banding in human chromosomes
XXXI   Centromeric heterochromatin in somatic chromosomes of (i) man, and (ii) meiotic chromosomes of *Nigella damascena*

FIGURES

1 *Drosophila* Larva   *page* 56
2 Metal Instruments   141
3 Glass and Earthenware Implements   142
4 Glass Bottles and Lamp   143

TABLES

I	Approximate Capacities of Apochromatic Objectives	*page*	24
II	Refractive Indices (*n*) of Mounting Media, etc.		48
III	Breakage of Chromosomes by X-rays		69
IV	Breakage of Chromosomes by Other Radiations		70
V	Chemical and Other Non-radiation Breakage of Chromosomes		72
VI	Control of Mitosis by Chemical Agents		74
VII	Physical Properties of Filters		89
VIII	Scope of Autoradiography		96
IX	Chromosomes of Natural Groups		102
X	Chromosome Calendar		107
XI	Properties of Reagents		111
XII	Composition of Compound Fixatives		113
XIII	Maceration Methods		125

*The co-existence of the most wonderful success
with the most profound ignorance is one of the
characteristic features of present-day biology.*

Szent-Györgi (1940)

1

Origin, Scope and Purpose of Chromosome Work

In 1831 Robert Brown discovered and named the cell nucleus. He saw that it was important, but he did not know what it did or how it worked. He could not at that time realise the part it played in cells and in organisms and even in the whole unfolding of life. These things we have since learned, and our knowledge may be framed in one sentence: *the nucleus is the chromosomes*. In the nucleus the apparatus of cell-government is at rest, in the chromosomes it is in movement.

But it was fifty years before the chromosomes themselves were named. Their study sprang from the study of whole organisms, tissues and cells. Its methods and its ideas were at first derived from its larger scale origin. The subdivision of its teaching between botany and zoology and medicine still lives on to remind of this accident of birth. And the anatomist from long habit still attaches to them a value (as he believes) consistent with their size. As Robert Brown said: 'The few indications of the presence of this nucleus, or areola, that I have hitherto met with in the publications of botanists, are chiefly in some figures of epidermis. . . . But so little importance seems to be attached to it, that the appearance is not always referred to in the explanations of the figures in which it is represented' (1833, p. 713).

Now, however, there is less danger of the nucleus being overlooked, and the chromosomes have become a study in themselves; a study, that is, with its own theory and its own technique. They are among the largest molecules whose chemistry is pursued. They are also the smallest living structures whose movements and transformations can be seen and followed. And throughout living organisms they show a uniformity in chemistry and mechanics which we now know to be the foundation of their uniformity in physiology and genetics.

The similarities of mitosis in the simple filaments of *Chara* and in the staminal hairs of *Tradescantia*, of meiosis in the turbellarian testis and in the paeony anthers, of the nucleic acid cycle and its vagaries in maize and in the fruit fly – these things may still astonish us. But now we have also learned to profit by them. They show us that we may turn from one to another in seeking material for experiment or demonstration. And in doing so we need consider only our own technical convenience. The lily can tell us what happens in the mouse. The fly can tell us what happens in man, though in certain circumstances, when necessary, the chromosomes of man can now more often be studied in their own right. We can use whichever is largest, easiest, best known or best suited, or merely what is in season.

This unity in the study of chromosomes arises from the uniformity of their chemical character. All chromosomes are macromolecules of nucleoprotein which have a precisely uniform and definable element in their structure. This element is DNA or desoxyribose nucleic acid. In all chromosomes of all organisms it is composed, we believe, always of the same four units or nucleotides arranged merely in different orders and proportions. A second element, the protein, is less accurately definable, and a third element RNA or ribose nucleic acid, consisting of at least three distinct classes, which are formed on the DNA as a primary gene product and intimately involved in synthesis of most of the protein in the cell.

Our study of chromosomes therefore has three aims.

First, we have to examine the chromosomes as chemical structures within the cell manifesting laws of behaviour peculiar to themselves, and fundamental to the reproduction of the cell and the organism.

Secondly, we have to examine them as the governing organs of cell life; that is to say, organs whose change underlies the variability or the constancy, the health or the disease, of animal or plant development. It is this function which makes their control the business of cancer research.

Thirdly, we have to examine them as the gene-strings, bearers of heredity, whose laws their uniform movements express. It is this last function which has made their control necessary for the practice of plant and animal breeding.

These primary aims have thus beyond them an ultimate one, that of the control of life in its three aspects, of reproduction, heredity and development. This control, as we shall see, has to be exerted through the agency of chromosomes by way of four processes: (i) mitosis and the reproduction of genes, (ii) meiosis and their recombination, (iii) fertilisation and polyploidy, (iv) mutation, breakage and reunion of groups of genes. To have this power we must know all stages of chromosome life and know them in organisms chosen as most suitable for study, that is, once more, for their technical convenience.

Chromosome treatment over the years has advanced by innumerable steps and strides; equally in fixation and in staining, in crude handling and in fine optics. The history of *Drosophila* is a pretty and special example of a collateral succession of developments. Breeding, chromosome counts, X-rays, salivary glands, transplantation, these tricks and devices applied to this favoured fly, have broadened the basis of experiment and have finally joined genetics and cytology into one practice and one theory.

A union that will play an ever increasing part in human society in the prevention of disease and possible regulation of transmission of harmful genes from one generation to the next, with indeed great implications for the future of linkage studies and mapping of human chromosomes, now made feasible by the recent introduction of banding techniques.

Many of the chromosome methods that are most effective are also extremely simple. Others are as elaborate as you could wish. The student, the teacher and the research worker cannot at present get hold of the information they need from any single source, and much of what is needed has never been published. Reagent, technique, organism and problem are all bound up together, and their suitability for one another has to be set out in the light of what we know to-day.

Experience has shown, however, that techniques arise in two stages. First, there is the study of technique in its general character as applicable to all organisms. Secondly, there is the study of our needs in working with particular organisms, tissues, and stages of development and for particular purposes. We have, for example, various structures in the chromosome: centromeres, nucleolar organisers, heterochromatin and chiasmata. We have nuclei in cells with and without cell walls, in dividing cells, and in variously differentiated cells.

And we have a host of experimental treatments for altering the course of development or of evolution. All these secondary needs react on the primary techniques to give us the right method for each situation. We shall now describe the simple methods fully for those who need to use them, and indicate how, why and when the elaborations and refinements of specialists need to be applied.

2

Equipment

A. Microscopes

The standard microscope for chromosome work must include the following components:

(1) Two apochromatic objectives, (i) 16 mm dry and (ii) 2 mm or 1·5 mm oil immersion (*cf*. Table I).
(2) A condenser under the stage.
(3) A plane mirror or prism.
(4) Three eyepieces magnifying approximately 5, 15 and 30 times.
(5) A mechanical stage.
(6) A camera lucida.

Achromatic instead of apochromatic objectives can, however, be used for all but the most rigorous work.

In addition, an 8 mm objective and 10 and 20 eyepieces will be useful for most workers and indispensable for some. For photographic work a revolving stage is convenient in adjusting the direction of the negative in relation to the field.

The use of a double, or binocular, as opposed to a single eyepiece is of no value for correct observation, it entails a loss of illumination and again a great waste of time if frequent drawings and photographs are made. Its use seems to be in enabling workers to carry out routine observations such as have now become superfluous for most purposes with the present technique.

Three other kinds of microsope are required for special purposes:

(1) For determination of chromosomal sites giving fluorescence with acridine derivatives: a fluorescence microscope, preferably with maximum aperture oil immersion objectives and fitted with a gas discharge ultra-high pressure mercury lamp and suitable filters.

(2) For watching and regulating the differentiation after staining: a low-power microscope with a concave mirror instead of a condenser and with 8 mm and 16 mm objectives, which need not be apochromatic. Daylight can be used for illumination.

(3) For the dissection of embryo-sacs, testes and salivary glands and other special tissues: a Greenough binocular microscope. An opal electric lamp is used for illumination.

B. Resolving Power and Magnification

Resolving power in the microscope is its capacity of producing a separate image of different parts of an object measured by their distance apart. It is limited by the angle of the cone of light illuminating the object and passing through the objective. Magnification on the other hand consists in the multiplication of the angle subtended by the image at the eye relative to that subtended by the object. Magnification, which derives from objective, tube-length and eyepiece (apart from resolving power) is necessary for providing an image large enough for the observer's eye to resolve and record its details. It is valueless as a means of increasing resolving power, and the highest eyepieces are wasted unless used with a system providing the appropriate resolution. Excess of either magnification or resolution relative to the other can be spoken of as empty.

The observer's eye escapes the theory of pure microscopy in two ways. First, observers differ in their capacity of resolution. What is empty magnification for one may be necessary for another. Secondly, the resolving power of the eye varies as between different wavelengths of light. Coles (1921) says that although the light best resolved by the microscope is blue, green is best resolved by the eye.

But it may be argued that resolution by the eye is greatest with white light, for then all the cones (red, green and blue) come into action.

The maximum resolving power is obtained with (i) critical illumination, (ii) an oil-immersion achromatic condenser of numerical aperture (n.a.) 1·4, (iii) an oil-immersion apochromatic objective of the same n.a. Whichever of these two has the lower n.a. necessarily limits the effective cone of light, and hence the resolving power of the whole system.

Most accounts of microscopy treat the microscope as an ideal optical instrument. It can however be treated as an instrument with two degrees of optical efficiency. The lower alone is necessary for routine examination, elementary teaching, and even for research on large chromosomes which are far above the limits of microscopic resolution. The higher degree is required for smaller scale examination and for photography. Some of the precautions necessary for the second are tedious and wasteful if applied to the first.

For most routine studies resolution up to 0·37 μm is sufficient, it is un-

TABLE 1 – APPROXIMATE CAPACITIES OF APOCHROMATIC OBJECTIVES

(For use with compensating eyepieces only)

Focal Length	Numerical Aperture n.a.	Magnification (with) 160 mm tube length $n = \dfrac{t.l.}{f}$	Limit of Resolution* $h = \dfrac{0.61\,\lambda}{n.a.}$	Depth of Focus† df	Object-Objective Distance‡
dry					
$\frac{2}{3}''$–16 mm	0·25	10	1·51 μm	12·2 μm	8·0 mm
$\frac{1}{6}''$–4 mm	0·75	40	0·50 μm	1·3 μm	0·7 mm
oil					
$\frac{1}{8}''$–3 mm	0·90	53	0·44 μm	1·0 μm	0·47 mm
„ –2 mm	1·3	80	0·29 μm	0·4 μm	0·37 mm
„ –2 mm	1·4	80	0·27 μm	0·4 μm	0·22 mm
„ –1·5 mm	1·3	107	0·29 μm	0·4 μm	0·25 mm

(*) Calculated for green light, $\lambda = 0.54$ μm.

(†) In medium of refractive index 1·5 (from Martin and Johnson 1931).

(‡) Subtract about 0·17 mm to give 'working distance' above the cover slip.

Note.—Cover slip thicknesses are as follows: No. 0, 75–100 μm; No. 1, 100–167 μm; No. 2, 167–215 μm.

necessary to go down to 0·27 μm. Since oil on the condenser is always trouble-some, a dry condenser can be used. This automatically limits its n.a. to 1·0. The whole system, with a 1·3 n.a. objective, will have an effective n.a. of 1·15 (*cf.* Coles 1921, Chamot and Mason 1938) and the limit of resolution will be about 0·33 μm. If it is never intended to use oil then an aplanatic condenser of n.a. 1·2 will be enough. With any condenser water may be used instead of oil as a compromise between rigour and convenience (Belling 1930).

C. Lighting and Screens

In practice the limiting factor in the efficiency of microscopes used for chromosome work is not usually the miscroscope but the lighting system. The simplest method of lighting for low powers is by a white sky with plane mirror and with the condenser focusing the beam at the level of the object on the slide and centred in the axis of the tube. The intensity of illumination produced in this way is of course uncontrolled and its source is diffuse. It is the business of artificial lighting to produce an image of the light source that is at once precise, constant and undiffused, and focused on the object; this constitutes critical illumination. The source must be small and uniform. If it shows any structure the condenser has to be put out of focus, with a slight loss of resolving power. Its radiation must be condensed by a convex lens into a nearly parallel beam filling the whole substage con-denser.

Two filter systems are necessary in fluorescence microscopy. A primary or 'excitor' filter is required between the light source and specimen to ex-clude all but the light of wavelength needed to stimulate emission of fluorescence in the fluorochrome being used, while a secondary or 'barrier' filter must be inserted between the specimen and eye, usually in the eye-piece, so as to ensure that only the light from the fluorescing specimen reaches the eye. The choice of filters in correct combination is dependent both on the fluorochrome being used and the spectral characteristics of the light source. Excitor filters 4 mm BG-38 or 1·5 mm BG-12, used in combina-tion with a K-530 nm barrier filter and a HBO-200 mercury lamp light source are suitable for the fluorochrome quinacrine mustard or its dihydro-chloride.

If the intensity of illumination is not adjustable at the source it may be reduced by any of four methods:

(1) reducing the source diaphragm,
(2) reducing the substage diaphragm,
(3) putting the substage condenser out of focus,
(4) inserting coloured screens, either substage or in front of the light.

For high powers, the first may already be reduced to a minimum needed for critical illumination. The second cuts down the cone of light, and so reduces the resolving power of the objective. The third method is only to be recommended for routine examination of under-stained or faded preparations with low eyepieces. The effect is produced by using the differential refractive index of cell walls and chromosomes to show them up. It is thus a kind of improvised dark-ground illumination (*q.v.*).

The last method is the best available for critical work. The choice of screens for eye work is governed by two considerations, the colour of the light and the colour of the stain. In general a weak screen should correct any colour in the light. A strong screen should be complementary to the stain: a blue screen for fuchsin and carmine, a yellow-green screen for crystal violet (*e.g.* Wratten filter 58 with a maximum transmission at 5200 Å). A merely reducing effect can be obtained by using a neutral grey screen. The loss of critical illumination from using ground glass in place of a screen is allowable for routine purposes.

A plane glass mirror is normally used for reflecting the beam of light into the axis of the microscope. Such a mirror produces slight secondary images of the source. To avoid these an aluminised mirror or prism can be used for reflection.

D. Oil and Lenses

Immersion oil should have a refractive index as close as possible to that of Crown Glass, *i.e.* 1·518 (*cf. Mounting Media*, Ch. 7). Cedarwood oil has been used for this reason in the past but it dries on exposure and needs to be cleaned off with lens paper moistened with xylol. Now a non-drying mineral or vegetable oil mixed with α-bromonaphthalene ($n = 1·660$) can be used of such a composition as the following:

(i) Liquid Paraffin 75·1%; α-Bromonaphthalene 24·9%.
(ii) Olive Oil 75·9%; α-Bromonaphthalene 24·1%.

Since refrective index may change with age the proportions may be corrected by testing with a refractometer. Proprietary oils should be tested in the same way.

These non-drying oils can be removed without xylol and also without direct pressure on the face of the lens. If finger-tip pressure is applied to the conical mounting alone, the lens will be cleaned well enough. This is important because it is impossible to ensure that, with the safest keeping, lens paper will be entirely free from grit. A non-fluorescing immersion oil should be used with the fluorescence microscope.

Dry lenses should be dusted with a camel-hair brush, cleaned with xylol.

Eyepieces and dry objectives can be dismounted for this purpose. Immersion lenses should be dismounted only by an expert.

E. Camera Lucida

For accurate drawing it is necessary to impose on the microscope image an image of the drawing paper of nearly equal intensity. This is done by the combination of a mirror and prism with an adjustable set of screens for reducing the strength of each beam to produce equality. A bench lamp beside the microscope is useful to maintain uniform lighting.

The Bristol board must be pinned on the drawing board, which must lie in a plane parallel to that of the microscope stage. A departure from this plane of 5° or 10° leads to a slight exaggeration of length; cos 5° is 0·996 and cos 10° is 0·985. A greater exaggeration of breadth, and one which is unequal for different parts of the drawing, results from the camera lucida mirror not being placed at 45°; for then the different parts of the drawing will not be at a distance from the eye proportionate to the distances of the corresponding parts of the image. The design of the camera lucida allows of this false adjustment and its convenience is a trap to be shy of, since measurement is one of the chief objects of drawing.

To obtain an accurate measurement of the magnification of drawings the scale of a *stage-micrometer* should be drawn under the same conditions as those of the chromosome drawings.

For counting pollen grains and recording their germination a squared *micrometer-eyepiece* attachment is also necessary.

F. Other Equipment

A list of other equipment is given in Appendix IV. One word of comment is needed. Workmanlike handling of these implements depends first and last on being clean, neat and orderly. Slides must be clean, bottles must be stoppered, implements in the right order. At every stage we work by wit and not by witchcraft.

3

Living Chromosomes

A. Material

The observation of chromosomes in the living state is useful as an introduction to their study, and it also serves certain special purposes of physical research. The optical methods which were formerly of interest for showing that living chromosomes tallied with the appearance of fixed ones are now chiefly used for the study of fixed material.

The most suitable material for direct observation of cell division and chromosome movement is as follows (*cf.* Becker 1938):

Plant:
  Staminal hairs and young petals, *Tradescantia,* diploid or tetraploid species. Belar 1929, Barber 1939, Kuwada and Nakamura 1940.
  Pollen tubes, many plants. Wulff 1934.
  Antheridial filaments, *Chara* and *Nitella.* Karling 1928; and rhizoids, *cf.* Becker 1938.
  Endosperm, *Iris, Leucojum, Haemanthus,* etc. Bajer and Bajer 1954, 1956; Östergren and Bajer 1958.

Animal:
  Salivary glands, Diptera, *Chironomus.* Bauer 1934; Poulson and Metz 1938.
  Testes, Orthoptera (Meiosis). Belar 1929, Nicklas 1967, Nicklas and Staehly 1967.
  Tails of Urodela. Barber and Callan 1942.

B. Methods

Removal of excess endosperm juices in the Bajers' method allows mild flattening of the cells by surface tension, the flattened cells entering mitosis in a normal way (Bajer, 1955; Östergren and Bajer, 1958).

The mounting media for these tissues are in order of importance:

(1) Liquid paraffin (Schaede 1930).
(2) Natural juices squeezed out of adjoining tissue.

(3) Ringer or isotonic salt solution.

(4) Sucrose solutions (especially for pollen tubes, *q.v.*).

(5) Agar with sucrose (Bajer and Bajer 1954).

Liquid paraffin may be used to ring the cover-slip and prevent evaporation even where it is not used for mounting. Its value depends on its capacity for dissolving five times as much oxygen as water will. In a class by itself is the examination of salivary glands in whole living larvae of *Drosophila* or `Sciara* in air or young *Chironomus* in water, ringed with vaseline to prevent squashing, and lightly pressed to show the gene bands (Buck and Boche 1938).

To exclude heat radiation from the source of light the beam should be passed through a water bath containing a weak aqueous solution of ferrous ammonium sulphate (Belar 1929*a*).

C. Vital Staining

The alleged vital staining of chromosomes by methylene blue is perhaps staining but scarcely vital, since this oxidising agent certainly injures the chromosomes as soon as it becomes effective. The same applies to a number of other vital stains that are said to have been effective (Yamaha and Nomura 1939).

D. Chromosome Optics

The chromosomes have three optical properties that make their observation possible without staining.

(1) For the first method of observation it is necessary to cut down the lighting so as to make use of the *differential refractivity* of the chromosomes and the spindle. The simplest method, used by Kuwada and Nakamura 1940, Wada 1932, is to restrict the cone of light passing through the condenser. Films of chromosome movement can be taken in this way (Barber 1939). For the study of the action of fixatives dark ground illumination has been found useful (Strangeways and Canti 1927). The principle of this method is to illuminate the object from the edge of the normal cone so that the light entering the objective is chiefly that transmitted by differential refraction. It is not therefore reliable as a means of studying fine structure.

(2) The second method depends on their *differential absorption of ultraviolet* light. With an ultraviolet microscope, photographs of living chromosomes have accordingly been taken (Lucas and Stark 1931). The maximum absorption was found by Caspersson (1940) to be at 2600 Å. The greatest value of this method lies in the higher resolving power of light with the

shorter wavelength (*cf.* Table I) which then can be applied to fixed chromosomes, stained or unstained (Ellenhorn *et al.* 1935).

(3) The third method is that of *birefringence* shown by polarised light. It depends on the regular molecular orientation of the chromosomes especially when stretched naturally as in the filamentous grasshopper sperms, or artificially as in accidents of smearing. This property, which they share with the mitotic spindle, makes it possible for them to be seen between crossed Nicol's prisms by polarised light (Schmidt 1937). For this purpose, although living observation is possible, fixation in alcohol has been used since it is naturally more convenient for photographic purposes, and desiccation perhaps intensifies molecular orientation (Nakamura 1937, *cf.* Caspersson 1941).

All these three properties, like those of differential staining, are presumably due to the nucleic acid attached to the protein chain of the chromosome.

(4) The fourth method is by phase-retardation of light according to Zernike's phase-contrast method (Michel 1941, Burch and Stock 1942, Richards 1944, Barer 1955). The insertion of phase-changing plates into the optical system of the microscope gives visible differences in the intensity of light absorption between cell constituents having different refractive indices. Under these conditions the chromosomes from late prophase onwards appear opaque, indeed as opaque as if deeply stained.

(5) The fifth method is by Nomarski differential–interference contrast microscopy (Nomarski 1955), which because the numerical aperture of the condenser can be fully utilised gives superior resolution to phase-contrast. Further, the image is not degraded by the halo effect. Fine details such as spindle fibres can be resolved in living cells (Bajer and Allen 1966). Another advantage is that normal bright-field objectives can be used.

(6) The sixth method is by interference microscopy, in which the mutually interfering beams which produce the contrast are obtained by an interferometer system incorporated in the microscope itself. Again resolution is better than with phase-contrast, since the aperture of the objective is unrestricted by a phase plate and the image undegraded by the halo effect.

Another important application of the interferometer microscope is in the determination of the dry mass of living cells. The thickness and volume of cells, and the concentration of solids and of water can also be deduced in certain instances (Davies *et al.* 1954, Davies 1958, Barer 1956, Hale 1958).

4

Bulk Fixation

A. Principles

The purpose of fixation is:

(1) *Coagulation* of the constituents of the cell without solution, disintegration, or disturbance of their internal structure or external spacing. The curd must be life-size and it must contain the water, which makes up nine-tenths of the cell.

(2) *Mordanting*, in a broad sense, to produce surface conditions in the fixed structures that will enable them to hold the particular stains intended for making them visible.

(3) *Toughening* the curd so that it will resist embedding, or *softening* it so that it will not resist squashing.

The objects to be treated have certain invariable properties. The chromosomes are threads composed of fibrous protein. These threads are the most resistant material in the cell, and they are further strengthened during mitosis by being spirally coiled and covered with a sheath of nucleic acid. It is this uniform coat which makes the chromosomes stain with the same basic dyes in all organisms. Acetic acid swells them slightly, chromic and osmic acids contract them, but none of these reagents causes displacements.

The chromosomes themselves, therefore, present little difficulty and little variation in fixation. They exist, however, under various conditions of cell and tissue environment. The cell environment is of two kinds: (i) The spindle and cytoplasm which, though more watery than the chromosomes, nevertheless support them well under all but the worst conditions of fixation. (ii) The prophase nucleus, which is particularly fluid in the prophase of meiosis. Its coagulation without displacement by shrinkage requires rapid fixation and careful handling.

The tissue environment is of greater importance in plants than in animals for here the cell walls, and also hairs and wax on the epidermal surface, are an obstacle to mass fixation. They demand a special treatment for rapid penetration.

It is therefore in relation to the cytoplasm and the cell walls, on the one hand, and the after-treatment, especially the staining, on the other hand, that we have to consider the choice of fixing reagents for chromosomes.

B. Practice

Coagulation can be brought about by boiling water, dilute sulphuric acid, absolute alcohol, and a hundred other agencies. The boiling water uncoils spiral chromosomes, and either water or alcohol will cause collapse of prophase nuclei. Single cells can even be fixed by drying. Smeared pollen mother cells, for example, can be fixed in a desiccator. In this way separate punctured eggs of worms have shown excellent fixation (Ch. 8), and pollen grains from herbarium sheets have given satisfactory pictures of the nuclear conditions (Ch. 10).

Coagulation without disturbance can be best brought about by two single reagents: 45% acetic acid as a fluid and osmic acid as a vapour.

Acetic acid, with its small ions, penetrates a tissue rapidly, but it swells the protoplasm and does not toughen it for later treatment, or at least not quickly enough. This difficuly may be surmounted either by combining the stain with the acid for joint action, or by combining alcohol with the acid to fix and harden the protein, other than the chromosomes, in the cells so that the chromosomes may be contained and preserved for later staining.

Osmic acid vapour gives the best possible fixation, but without deep penetration, or indeed any penetration of cellulose walls. It can therefore be used only on animal cells, which should be smeared in a thin film and inverted over a chamber containing the 2% solution for half a minute. Since the preparation must afterwards be stained and therefore subjected to either dehydration or maceration, it must be toughened after fixation. This can be done with 1% *chromic acid* for 1 hour.

The combination of all these three acids has the necessary properties of penetration, coagulation and toughening. It was the foundation of the first effective fluid for fixation in bulk. This combination in weak concentration was first used by Flemming in 1882. Its variations have given rise to all the chief fixatives in use today. These variations have consisted in:

(1) Changes of concentration, *e.g.* strong Flemming, etc.

(2) Reduction in the acetic component, to preserve chondriosomes from solution, *e.g.* Benda, or for smear preparations where penetration is less important, *e.g.* La Cour's 2BE. The complete omission of acetic acid by Champy and Lewitsky is unreliable for plants on account of the cell walls, but Minouchi, Matthey and Koller have found it useful for mammals.

(3) Replacement of osmic acid by less expensive reagents. This was done completely by Bouin in 1896 with picric acid for animal material and by S. Navashin in 1910 with formaldehyde for plant material. It has been done partially in La Cour's series of fixatives with potassium dichromate. Picric acid, as well as other substitutions by platinic chloride, ruthenic acid, uranic acid and uranium salts we cannot recommend.

(4) Addition of reagents to reduce surface tension, or to raise the osmotic pressure, as discussed later.

A second line of development arose from the penetrating alcoholic fixatives first used by Carnoy in 1884. Their disadvantage lies in their unsuitability for embedding afterwards.

The aqueous fixatives harden a tissue for embedding. The alcoholic fixatives containing acetic acid soften cell walls and matrices conveniently for squashing, and by their lower surface tension allow of the penetration of resistant membranes such as egg shells.

Fixatives containing both oxidising and reducing agents, such as the Navashin and Bouin series, must have the opposite components kept separate until they are used, and should be used immediately on mixing. It is also said that the Champy series should be fresh, on account, perhaps, of its high osmic content.

C. Mordanting and Hydrolysis

The essential step in the fixation–staining process is in general one of salt formation of a non-specific character (Gulick 1941), and Feulgen's method of staining, which will be specially considered, gives a solitary example of a known chemical reaction being concerned.

In the salting process combination of stain and fixative is vital. Oxidising agents, as well as strong acids, prepare the surface of chromosomes and other bodies for particular staining reactions. Thus a formalin fixative probably gives a weaker nucleic acid reaction with crystal violet than does an osmic fixative. This difference shows only in nucleoli, the chromosomes being saturated after either fixation. Chromic acid, or some equally penetrating oxidising agent, is a necessary component of any fixative for crystal violet. On the other hand it interferes with the action of any acetic stain.

With some stains, however, no fixation can provide a sufficient mordant for staining. Mordanting must be carried out afterwards. The function of the mordant is either as an isoelectric point modifier or as a chemical link between the stain and its recipient. Possibly chromic acid before crystal violet and alum before haematoxylin has the first effect, while hydrolysis before Feulgen staining is known to have the second effect. An after-mordant effect is produced by rinsing a slide in iodine or in chromic or picric acids after staining with crystal violet.

D. Penetration

Rapid penetration of the fixative is necessary for its uniform action. This is aided in six ways:

(1) Dissecting out small testes (under Ringer-solution) or anthers, cutting up under the fixative, slitting root-tips or truncating anthers, or (see Ch. 5) smearing material before fixation. In addition:

(i) Animal eggs (Ch. 8) may be pricked to admit fixative.
(ii) Pollen mother cells which hang together and are thus impossible to smear (*e.g.* in some lilies and cereals) can be squeezed out of cut anthers in four strings and fixed naked for embedding.

(2) Penetration of the surface by alcoholic fixatives is most satisfactory, but they fail to harden the important cell structures. To remedy this they can be followed after one minute by rinsing and the use of an aqueous fixative (Kihara's Carnoy–Flemming method). This method is useful where flower-buds have to be fixed whole.

(3) Some fixatives, *e.g.* Allen's Bouin or acetic acid, can be used warm (36°–40° C). This increases the rapidity of penetration and it may increase the effectiveness of coagulation and hardening.

(4) After unheated aqueous fixation (but never after alcoholic treatment) the material should immediately be shaken and put under a water or hand vacuum pump to remove air bubbles which may delay penetration.

(5) Materials may be added to an aqueous fixative to lower its surface tension. This is the purpose of saponin in La Cour's fixatives.

(6) Maltose and urea have been used to adjust the osmotic pressure of fixation. The 'effective' osmotic pressure is, however, that due to small ions in solution (*cf.* Young 1935) and these substances are relatively useless. Adjustment of the effective osmotic pressure to that of the tissue to be fixed is important in bulk fixations of animal material and can be achieved by making up the fixing solutions in normal saline.

E. Condition of the Material

Material before fixation must be healthy. Starvation with animals, drought with plants, have a disastrous effect on the fixation properties of both chromosomes and mitotic spindle. Root tips for mitosis are best either taken from pot plants where the ball of soil is neither too wet nor too dry, or from water-cultures. Testes for meiosis should usually be from immature or young animals. Cold treatment is a palliative for poor condition in hot weather, and even the use of fixatives at freezing point.

F. Duration

Owing to the absence of cellulose walls, animal tissues are fixed more rapidly than those of plants. The hardening process, in both cases, requires several hours. The upper limit varies with the reagent as follows:

Fixative Type	Limit of Safe Action	Effect of Excess
Formalin	several weeks	—
Low Osmic	1 week	—
Alcoholic	24 hours	reveals spindle structure
High Osmic	1 hour	damages chromosomes

Since animal cells are not hardened enough in one hour, their fixation in high-osmic types must be followed by 12–24 hours in 1% chromic acid.

G. Uses of Bulk Fixation

Bulk fixation was formerly used for preparing the material for embedding and sectioning by the paraffin method (Ch. 6). Now it can also be used for the direct preparation of squashes (Ch. 5).

5

Smears and Squashes

A. Material

Sections have now been largely replaced by smears and squashes for all but the smallest masses of material. The advantages of these methods are rapidity of fixation and rapidity of handling without embedding. In both respects rapidity means not only saving of time but also increased efficiency. Both methods enable us to examine single layers of large cells in their totality. This advantage is combined with the instantaneous fixation of smears and the rapid acetic fixation of squashes.

B. Smear Methods

Where feasible, smearing is to be preferred to squashing for meiosis in anthers and soft testes. Smearing consists always (in the absence of pretreatment) of the direct spreading of the cells in such a semi-fluid tissue over the surface of the slide with a flat-honed scalpel and the immediate inversion of the slide over a dish of fixative. The combinations of fixative and staining that are possible with smearing are shown in Schedule 2.

C. Acetic Squash Methods

The simplest of all chromosome treatments is to use a combined stain-fixative of the aceto-carmine type first used by Schneider and adapted for chromosome work by Belling in 1921. For this purpose inversion is unnecessary, since the cells can be directly mounted and studied in the fixative. This makes acetic staining methods very convenient for determining stages of division in anthers and testes before fixing them by some other method.

The standard iron-aceto-carmine method was devised by Belling (1926) for the study of pollen mother cells, but it is now equally widely used for animals. It consists in teasing out the tissue in aceto-carmine with unplated iron needles and mounting directly under a cover slip. The iron acetate formed acts as a mordant for the carmine and gives a differentiation of chromosomes and cytoplasm which is usually good enough for the study of metaphases.

In many flowering plants, however, particularly where the chromosomes are small, the cytoplasm takes up too much of the stain. The following improvements and developments of the aceto-carmine technique are chiefly concerned with this difficulty.

(1) *Fixation.* A prior fixation in 1:3 fresh acetic alcohol for 12–24 hours reduces the staining of the cytoplasm (McClintock 1929). Whether this effect is due to differential solution or change of pH is not known. Propionic acid can be used to replace acetic acid with advantage for some materials.

(2) *Storage of Material.* If the material is not wanted for immediate staining, it can be stored in 70% alcohol at 0–4°C, but fixation and staining will suffer after two months or more. Long storage however can be used for the special purpose of exaggerating the differentiation of the spindle to facilitate the study of its development (Darlington and Thomas 1937).

(3) *Spreading.* Gentle heating of the slide over a spirit flame flattens the cells, sticks them to the slide and cover slip and spreads the chromosomes whether in prophase or metaphase (McClintock 1929). The degree of spreading is due to the pressure of the cover slip, and this is controlled partly by the amount of material on the slide, partly by the amount of stain-fixative used and partly by pressure on the cover slip. All useless debris should be removed, preferably with the aid of a dissecting microscope. This is essential for the flattening and spreading of the chromosomes at pachytene. Gentle finger tip pressure on the cover slip under filter paper may assist in this process, but lateral movement of the cover slip should be avoided if entire cells are required. A minimum-sized drop of stain-fixative, compatible with allowing the placing of the cover slip in position without air bubbles, should be used. To obtain even staining 30 to 60 seconds should elapse before applying the cover slip.

(4) *Pre- and Post-mordanting.* The method of Thomas (1940) is useful for small chromosomes. Iron acetate is introduced into the acetic alcohol. A weak stain is best, about 1/3 strength aceto-carmine (diluted with 45% acetic acid) and more iron is introduced at the time of teasing by using steel needles.

Alternatively a trace of ferric chloride may be added to the acetic alcohol (a few drops of concentrated aqueous solution) to give the fluid a pale straw colour. This method leaves the cytoplasm in a soft and plastic condition facilitating easy flattening.

Iron alum can be used as a mordant *after* fixation. It is a most satisfactory method for Algae (Godward 1948) and for pachytene in the tomato when followed by differentiation in 45% acetic acid over a steam bath (Barton

1950). Prolonged staining in alcoholic hydrochloric acid-carmine may prove advantageous for some material (Snow 1963).

(5) *Acetic-Orcein.* This chromosome stain recommended by La Cour (1941) is more selective than carmine and for many organisms gives clearer permanent preparations. No iron mordant is needed. It can also be used in 45% propionic acid with advantage for some species.

It is now the recognised stain for salivary gland chromosomes of Diptera (Ch. 8, *d*. Schedule 4).

Acetic-orcein has also proved valuable for the chromosomes of mosses (Vaarama 1949). In some tissues and organisms, the selectivity of staining can be further improved if, after acetic alcohol fixation, the tissues are hydrolysed for 5–8 min in N HCl at 60° C before staining. This has proved advantageous in various species of Ascomycetes (*cf.* Singleton 1953, Day *et al*. 1956, Elliott 1956).

(6) *Lacmoid Indicator* (resorcin blue). This reagent was first used in alkaline solution for staining cytoplasm (*v*. pollen tube technique). In acetic acid solution it stains only the chromosomes. Again no iron mordant is needed. Its colour remains red so long as the preparation remains acid and the mount is made in cedarwood oil. If the mount is made in neutral balsam the colour changes to blue. Lacmoid is most useful after maceration (see below).

(7) *Bismarck Brown.* Formerly used as an aqueous stain this reagent can likewise be substituted for carmine (Plate XII, 1).

(8) *Compound Staining.* Nebel (1940) has suggested the use of chlorazol black E as a combined stain with carmine for small chromosomes. Workable stains can also be made from all combinations of orcein, lacmoid and carmine.

(9) *Storage of Slides.* Preparations are usually improved after keeping for a few days. The edges of the cover slip should be sealed with a rubber solution, or a mixture of gelatin and acetic acid (App. II). They are best kept in a refrigerator at 0–4° C.

(10) *Permanent Method.* There are a number of methods for making permanent preparations, the most notable being McClintock's (1929), the vapour method of Bridges (1935) and the dry ice method of Conger and Fairchild (1953). The last method should be avoided with carmine, for it is easier to avoid change, or correct for over-staining with other methods. A modified form of McClintock's method is given in the schedule.

Combined fixing, staining and mounting methods developed by Zirkle (1940) can be used with water-soluble or fat-soluble mounting media (App. III). The results are rapid and often satisfactory. Orcein and lacmoid can be used to replace the carmine in any of these formulae.

In conclusion we may say that no one type or method of acetic staining-fixation is best for all materials. On the whole orcein and lacmoid are better than carmine for all animal tissues. Both are excellent also for many plants, but may burst some pollen mother cells whose cytoplasm is delicate, *e.g.* *Allium*. Owing to its slight hardening effect, for pollen mother cells of plants with small chromosomes and dense cytoplasm, *e.g. Rosaceae*, aceto-carmine combined with premordanting is the most satisfactory. For some organisms and special tissues (see Schedule 4) orcein is better than either carmine or lacmoid.

D. Maceration

Soft fixations of soft tissues, *i.e.* alcoholic fixations of PMC or insect testes, need not, indeed should not, be macerated for squashing. All other material requires maceration, which must be so adjusted in plant tissues that solution of the pectic salt in the middle lamella of the cell walls is complete without the cell contents being unduly softened. There are eight methods of maceration:

(1) Add N HCl to acetic orcein or acetic lacmoid.

(2) Extend Feulgen hydrolysis to maximum time (see p. 46).

(3) Immerse in equal parts of conc. HCl and 95% alcohol for 5 min without warming. Further, after alcoholic fixation, harden the material in Carnoy for 10 min before staining (Warmke 1935).

(4) Fix in 5 N HCl for 15 min at 20° C (see Schedule 7, Marks 1973).

(5) Treat for 2–5 min in sat. aqu. ammonium oxalate and H_2O_2 (equal vols of each) but avoid tap-water for rinsing. Use when the fixative contains osmic acid (Ford unpub.). For meristematic tissues only.

(6) To assist maceration after osmic fixation, fix for 2 hr and then treat tissues overnight in equal parts 1% chromic acid and used fixative. If required the treatment can be followed by that in (5).

(7) Treat with snail stomach cytase overnight (Fabergé 1945) or 5% pectinase in 1% aqu. soln. peptone 4–16 hr (Chayen and Miles 1953). Use for all plant tissues except pollen grains.

(8) Treat PMC with large chromosomes for carmine staining with a fungal enzyme, clarase, 1% soln. for 10 min to 3 hr (Emsweller 1944).

Root Tips. All these methods except the last are suitable. There are some difficulties however in plants with numerous small chromosomes. Most of the metaphase plates are then found to be in side view, and the cells cannot be turned without damage. Pre-treatment with drugs (Ch. 9) can overcome this difficulty by inhibition of the spindle; metaphase plates with super-contracted chromosomes are then accumulated in the tissue. To avoid too much contraction O'Mara (1939) immerses excised roots in 0·01% colchicine for less than 3 hours. As an alternative α-bromonaphthalene in saturated solution is cheaper and as satisfactory for most species (O'Mara 1948). Low temperature can produce a like effect. Thus transference of *Trillium grandiflorum* from 24° C to 0° C for two days greatly increases the proportion of mitoses which are in metaphase without causing too much contraction (Darlington and La Cour 1945). Warmke's method (1946) of keeping cut root-tips for 2 hours at 0° C before fixation probably serves the same purpose.

These treatments can be followed by squash methods, but where super-contraction must be avoided, one of the following two methods may be tried :

(i) *a.* Carnoy fixation. *b.* Transverse sectioning by the freezing microtome.
    c. Combined softening and staining in acetic-lacmoid (Schedule 3).
(ii) *a.* Aqueous fixation. *b.* Transverse sectioning by the paraffin method.
    c. Feulgen staining with extended hydrolysis. *d.* Pressure applied after mounting in thin balsam (Warmke 1941).

Leaves. Where roots are unobtainable rapidly growing leaf buds (the leaf meristem) as well as petals, tendrils or glumes provide a valuable source of mitoses (Baldwin 1939, Darlington and Thomas 1941). Again to accumulate metaphases treatment of excised leaf tips in 0·2% colchicine, kept in light for 1–2 hours, is suggested by Meyer (1943). We find that the concentration of colchicine can be as low as 0·05% depending on the greenness of the leaves.

Pollen-Grains. Method (3) is the most suitable. Maceration allows the contents of the pollen-grain, whether large as in *Lilium* or small as in *Pyrus*, to escape from the thick wall which in these cases stands in the light of the chromosomes. An exine solvent (App. II) has been used on *Gossypium* pollen by Bernardo (1965).

Embryo-Sacs. Method (1) has a sufficiently hardening effect on the cytoplasm to preserve the embryo-sac entire.

6

Paraffin Methods

A. Value

Embedding may become necessary where, for one of several reasons, smears or squashes are impracticable. This is particularly so in the handling of very small organs, including all but the largest embryo-sacs, or even for larger bodies where mitoses are scarce. As an auxiliary method, embedding is also useful in showing the arrangement of cells in a tissue and the sequence of the stages of meiosis in a testis or an anther.

B. Fixation and Washing

For material that is to be embedded, aqueous fixatives are necessary to protect the cells from maceration by water or collapse by loss of water. The precautions already described must be taken to ensure satisfactory penetration. A solution of at least 20 times the volume of the material to be fixed must be used to fix it, and bulk material in any case requires stronger fixatives than do smears.

For most purposes long washing in water is unnecessary and even harmful. Indeed washing can be omitted altogether in preparing root tips for chromosome counts (Randolph 1935, Upcott and La Cour 1936). The root tips can be taken directly from the fixative into 70% alcohol. For more delicate tissues 2 or 3 changes of water are preferable before transference to alcohol.

C. Dehydration and Infiltration

Many attempts have been made to find reagents as suitable substitutes for ethyl alcohol and xylol, or ethyl alcohol and chloroform. The objects have been a speeding up of the process, an avoidance of excessive hardening and shrinkage of tissues, and the use of one fluid instead of two. The most important of these reagents are:

(1) *Dioxan:* Graupner and Weisberger 1931, 1933; Johansen 1940; Baird 1936 (animal tissues); La Cour 1937; Maheshwari 1939.

(2) *n-Butyl alcohol:* Zirkle 1930, Randolph 1935.

(3) *Tertiary butyl alcohol:* Johansen 1940.

(4) *Iso-propyl alcohol:* Bradbury 1931.

(5) *Methylal paraffin oil:* Dufrenoy 1935.

Of these reagents, the first three have given the most promising results. Dioxan, however, has toxic properties and extreme care should be taken in its use. Johansen claims that his method is the least drastic and does not remove all the bound water from the tissues. It should therefore avoid collapse of delicate structures such as prophase chromosomes. All substitutes for chloroform are nevertheless poorer solvents of paraffin wax and for the infiltration of bulky material, such as flower buds, chloroform is to be preferred.

Recent improvements, however, in the quality of ester waxes (Steedman 1960) now make it possible to avoid the reagents largely responsible for excessive hardening and shrinkage. Ester wax will tolerate up to 5% of water in solvents such as ethyl alcohol so tissues can be infiltrated with the wax direct from 95% alcohol if desired (Schedules 1A, 1A').

Special treatments are useful for small objects (Madge 1936). Eosin may be introduced in the 70% alcohol, or fuchsin in the 1:3 alcohol–chloroform, to make them conspicuous in the wax.

D. Embedding

The following precautions should be borne in mind:

(1) Remove all traces of the solvent by evaporation, or by changing the molten wax.

(2) Avoid overheating. The oven temperature should never be more than 60° C.

(3) Choose wax of melting point suited to the tissues to be cut and the laboratory conditions. In a temperate climate use 50° C m.p. wax for sections to be cut at 14–40 μm and 58° C m.p. below 14 μm. Waterman (1939) has described the preparation of hardened paraffin waxes having low melting points; ester waxes with these qualities can be obtained commercially.

(4) Orientate root tips and anthers in rows in the wax for convenience in cutting. Randolph (1940) has suggested a method of card mounting which might be valuable for small root tips. The tips are mounted prior to fixation. Fabergé and La Cour (1936) have devised an electrically heated needle to

facilitate orientation, avoiding the danger of damaging delicate tissues by overheating.

(5) Cool the block rapidly in water to avoid crystallisation of the wax.

E. Section Cutting

The first question to decide is the thickness at which the sections should be cut. Until recently, for heavy haematoxylin staining, sections had to be cut so thin that a large proportion of nuclei were injured. Indeed some microtomes are not adjusted to cut at more than 20 μm. The best extend to 40μm. The following list shows how to adjust the thickness of the sections to the length of the chromosomes:

Length of Chromosomes		1–10 µm	15–40 µm
Thickness	RT	4–16 µm	20–40 µm
of	PMC	14–20 µm	30–40 µm
	EMC	20–30 µm	40 µm
Sections	Testes and Eggs	4–16 µm	20–40 µm

(1) Keep the microtome knife sharp by correct honing and stropping. For a review of methods see Maheshwari (1939).

(2) To secure straight ribbons with a minimum of breaking, trim the block so that the wax is evenly distributed around the material. The sides must be parallel and mounted parallel to the razor edge.

(3) Mount the razor at the correct angle and screw it tight to keep it rigid.

(4) If there is difficulty in securing continuous ribbons of thick sections in cold weather, either warm the mounted block and holder in the oven, or warm the edge nearest the razor with a hot scalpel.

(5) For thin sections that wrinkle badly, owing to the wax having too low a melting point for the room temperature, cool the block on ice or use a hardened paraffin (see under embedding).

(6) To secure sections of hard material (animal eggs) expose one side of the material by trimming the block, and soak in water 12–24 h. Waddington and Kriebel (1935) advise the use of a small amount of petroleum ceresin in the embedding wax.

(7) In dry climates the ribbons will be electrified. Hance (1937) has suggested air conditioning to avoid this. For other methods see Maheshwari (1939).

F. Mounting Ribbons

Mayer's albumen is most commonly used for fixing ribbons to the slide. The film must be the thinnest obtainable by smearing with the finger. Even so, it will be well to avoid loss of thick sections through failure of drying, by heating the smeared slide gently over a spirit flame for 2–3 seconds.

The ribbons are cut into lengths shorter than the cover slip to allow for stretching.

The ribbons are floated on a few drops of water on the slide. A hot plate maintained at 45° C is necessary for stretching and drying the ribbons. Avoid air bubbles under the sections. They arise from careless laying out or too sudden heating. Drying is completed in 4 to 12 hours according to thickness. It can be hastened by the use of 20% alcohol instead of water.

G. Removing The Wax

Before the ribbons can be stained the wax must be dissolved in xylol and the slide then placed in absolute alcohol and after bleaching if necessary (Ch. 7) taken down to water. (Schedule 1.)

An excellent review of sectioning methods has been written by Steedman (1960).

Note. In Britain permits for duty-free absolute alcohol are obtained from a Customs and Excise Office.

7

Staining and Mounting

A. Staining

Apart from the acetic stains which act as combined stain–fixatives, nine basic dyes are suitable for aqueous staining of chromosomes, as follows:

(1) The *leuco-basic fuchsin* method was first developed by Feulgen and Rossenbeck (1924) as a microchemical test for the desoxyribose-nucleic acid (DNA) found in chromosomes (Gulick 1940). It depends on Schiff's aldehyde reaction, and gives a translucent but fairly permanent stain. The method consists of:

(i) Liberation of the aldehyde groups of the nucleic acid by mild hydrolysis in normal hydrochloric acid at 60° C.
(ii) Chemical reaction between the liberated aldehydes and the leuco-basic fuchsin, resulting in a violet coloration of the chromosomes.

Three types of substance in the cell are capable of giving the aldehyde reaction, (i) free *aldehydes* in lignified cell walls, (ii) DNA after hydrolysis, (iii) *polysaccharides* after oxidation with chromic acid (Bauer 1933): (i) and (iii) thus require no hydrolysis but (i) is confined to the cell wall and (iii) is limited to the effects of chromic acid fixation. A specific diagnosis of DNA acid in the cell should therefore always be possible by means of the Feulgen reaction. Indeed the concept of DNA constancy for the haploid chromosome set is partly based on the quantitative estimation of Feulgen values in nuclei by microphotometry (*cf*. Swift 1953, Vendrely and Vendrely 1956). It should be appreciated, however, that probably only a relatively constant fraction of the DNA in the chromosomes reacts to the Feulgen stain.

The method is of wide application. Tissues can be stained in bulk (see squash methods), or as sections or smears. It can be used after alcoholic fixation, and unlike most chromosome stains, it requires no differentiation.

The Feulgen method is subject to occasional failure or apparent failure. This failure or seemingly weak reactivity has been due to various causes:

(i) Diffuse distribution of DNA in extended chromosomes or expanded nucleus, *e.g.* lampbrush chromosomes or the vegetative nucleus in pollen grains.

(ii)   Incorrect timing of hydrolysis. As to the minimum and optimum times of hydrolysis, Hillary (1939), using nucleic acid impregnated agar blocks and four different types of fixatives, has shown that two types of hydrolysis curve are possible. These depend on the presence or absence of chromic acid in the fixative. With fixatives containing chromic acid, maximum staining follows hydrolysis at 60° C from 6 to 30 min; without chromic acid the maximum extends only from 4 to 8 min (*cf.* Bauer 1932). After storage of tissues in alcohol it is often necessary to extend hydrolysis times. Further, hydrolysis sometimes has to be extended, possibly because in some plants the tissues contain substances inhibiting it, *e.g.* in root tips of *Chrysanthemum* (Dowrick 1952). In some instances this may perhaps be due to interference by protein (Shinke *et al.* 1957).

(iii) Osmic acid unless carefully reduced by bleaching will sometimes interfere with the Feulgen reaction. This is particularly serious with small chromosomes. Incomplete bleaching of the nucleolus has its use, however (see Plate XII).

(iv) Formalin fixatives will lead to staining of the cytoplasm by the Feulgen reaction, if not washed out.

(v) Poor samples of basic fuchsin, which fail to give completely decolorised solutions, will lead to stained cytoplasm, and sometimes weakly stained chromosomes. Coleman (1938) has suggested decolorising carbon as a means of obtaining colourless solutions free from impurity (App. II).

(vi) Kasten (1960) has listed 28 other dyes of various colours which will give a Feulgen-type reaction following acid hydrolysis. By choosing dyes of contrasting colours, it is possible to intermingle cells from different treatments in the same squash preparation, in order to obtain comparative measurements in equally squashed cells (Savage 1966).

(2) The *crystal violet* method (Newton 1927, La Cour 1937) is simple and rapid. It is possible to obtain sections up to 40 μm in thickness with well-stained chromosomes and clear cytoplasm. Also the intensity of the stain is easily reduced by differentiation, an important factor in the study of large chromosomes at meiosis. A weak stain, 0·1%, recommended by Upcott and La Cour (1936), can be used for this purpose. Where the stain does not hold in small chromosomes the chromic acid method of La Cour (1937) or the picric acid method of Johansen (1932) can be used.

(3) *Haematoxylin* is now seldom used except where intense permanent staining is required. Particularly in plants, coloration of the cytoplasm is difficult to avoid, and this stain is therefore more easily applied to thin sections, or smears, of animal tissues. A more elaborate method than that given in Schedule 10 is recommended by Cole (1926).

(4) The *brazilin* method of Belling (1928) and Capinpin (1930) suffers from the same defects as haematoxylin, but it can be made to yield good results with pollen mother cells.

(5) *Celestin blue B* and *gallocyanin* can be used as simple nuclear stains, preferably as iron lakes which can be prepared by boiling 2–3 min in a 5% solution of iron alum (Proescher and Arkush 1928). No mordanting or differentiation is required.

(6) *Giemsa* (Gelei's modification) has been recommended by Belar as a satisfactory stain for animal chromosomes after fixation by osmic vapour. Giemsa has also been used for staining mitosis in germinating spores of bacteria after fixation for 3–5 min in osmic vapour and for an indefinite time in 70% alcohol followed by hydrolysis for 7 min in N HCl at 60° C (Robinow 1941).

(7) *Azure A* in the presence of SO_2 stains nuclei selectively after a short hydrolysis in N HCl at 60° C. Thionyl chloride has been used by De Lamater (1951) as a source of SO_2 in the use of this stain.

(8) *Toluidine blue* is an effective stain for chromosomes when used following hydrolysis in N HCl at 60° C, or 5 N HCl for 40 min at 20° C to remove ribose nucleic acid (RNA). When combined with dimethyl hydantoin formaldehyde resin, it provides an effective stain–mounting medium, particularly for spreads of mammalian chromosomes (Breckon and Evans 1969, Evans *et al*. 1972). It is most useful in autoradiography (stripping-film technique) for staining through the film after photographic processing (Pelc 1956). Stain for 30–60 seconds in a 0·05% aqueous solution. Used hot it is possible to stain nuclei, nucleoli and chromosomes effectively in 'Epon' resin sections as thin as 65–70 nm (Wells and La Cour 1971).

(9) *Methyl green* as a nuclear stain is perhaps best known for its use in combination with pyronin. As used by Kurnick (1950, *cf*. 1953) it is believed to be specific for polymerised DNA, but this has been questioned by Alfert (1952). It is an unstable dye and solutions have to be repeatedly extracted with chloroform before use, to remove methyl violet.

Light green and fast green are suitable counter-stains to use for the nucleoli after the chromosomes have been stained by the Feulgen method. The light-green method of Semmens and Bhaduri (1941) gives specific staining. A simple fast green method for slide or bulk material (0·1% aqueous solution for 2–12 hours) has been recommended by Hillary (1940) to be used after bleaching in SO_2 water and rinsing.

B. Mounting

Optically the soundest mounting medium is one whose refractive index lies between that of the glass slide and that of the tissues, which may be a little higher. Canada balsam in general has the advantage on this ground. It is

TABLE II – REFRACTIVE INDICES (n) OF MOUNTING MEDIA, ETC.

Medium	n	Medium	n
Distilled water	1·336	Crown glass	1·518
Sea water	1·343	Cedarwood oil, thick	1·520
Liquid paraffin	1·471	Balsam in xylol	1·524
Olive oil	1·473	Clove oil	1·533
Glycerine, concentrated	1·473	Balsam, dry	1·535
'Euparal'	1·483	Gum damar	1·542
Xylol	1·497	'Clarite X'	1·567
Cedarwood oil, thin	1·510	(= 'Nevillite No. 1')	

however acid, and causes certain stains to fade. To overcome this difficulty various neutral resins have been produced. Neutral balsam in sealed tubes and 'clarite X' may be recommended, but clarite cannot be used for aceto-carmine preparations, and becomes milky with alcohol.

Xylol is the usual solvent for such resins, but toluene dries more quickly (Groat 1939). Soft tissues or squashes are liable to collapse in these petroleum distillates. To avoid this we may avail ourselves of two other methods:

(1) *Feulgen, carmine or orcein preparations:* Mount directly from absolute alcohol in 'Euparal' or balsam dissolved in alcohol. If the alcohol contains any water the balsam will become cloudy. Warm on the hot plate to get rid of this. Dioxan and dioxan–balsam are recommended by Hillary (1941).

(2) *Lacmoid preparations:* Mount directly from absolute alcohol to cedarwood oil or 'Euparal' which has become acid with storage (App. III).

C. Fading

All stains can fade in certain circumstances, though what these circumstances are has not always been defined. The chief agents of fading seem to be (i) ultraviolet light (since it will happen rapidly from an arc lamp projector and gradually from daylight), (ii) acidity of the mounting medium, already referred to, and (iii) inclusion of reagents from earlier treatment, through careless mounting.

From all these causes crystal violet fades most easily (even in a few weeks), haematoxylin and carmine least easily. F. H. Smith (1934) claims that washing in saturated picric acid in absolute alcohol after staining makes crystal violet permanent. His method certainly delays fading. The chromic acid after-mordant would probably have the same effect.

D. Re-staining

After fading or unsuccessful staining re-staining is possible with the same stain or with any other stain for which the original fixative was appropriate, *e.g.* a Flemming-crystal violet preparation can be re-stained in Feulgen. This is done by repeating the process after removing the cover slip and the mounting medium with xylol. To this rule there is one physical exception: an acetic smear is likely to collapse on removing the cover slip. And there is one chemical exception: Feulgen preparations cannot be re-hydrolysed for re-staining with leuco-basic fuchsin. They can however be re-stained, after acetic fixatives with an acetic stain (especially lacmoid), after aqueous fixatives with crystal violet and after alcoholic, with haematoxylin. Further, after acetic fixatives an ordinary Feulgen preparation which has turned out to be insufficiently stained may at once be brightened with orcein or carmine (2–5 min).

E. Bleaching

Before staining, sections and smears should be bleached in hydrogen peroxide where necessary (App. III). Bleaching is required to oxidise the black precipitate produced by fats with fixatives containing osmic acid. With bulk material for Feulgen squashes bleaching is usually not necessary. Hydrolysis removes part of the precipitate and squashing makes the remainder negligible.

8

Special Treatments

A. Pre-Treatment for Structure

Internal spirals, which appeared at metaphase by accidents of fixation, were first illustrated in *Tradescantia* by Baranetzky in 1880. Relic spirals have long been known in the resting and prophase nucleus. For a long time a relationship of the two was vaguely presumed. The general laws of chromosome structure were clarified only by the regular demonstration of internal spirals by pre-treatment of pollen mother cells (or testes in the few animals examined). This method derives from Sakamura (1927) who photographed major spirals in MI chromosomes of *Tradescantia* fixed in boiling water and unstained. A second step was the discovery by various workers that smears dried before chromic fixation (accidentally or otherwise) showed spiral structure. Diploid and tetraploid species of *Tradescantia* and many species of *Trillium* are convenient for the smear treatment of PMC necessary with all methods.

Some of the methods, such as heat and ammonia vapour, depend on uncoiling and can restore a resting nucleus condition. Others possibly depend on dissolving the histone associated with the DNA backbone.

A third effect is swelling, which depends on increased hydration following a change in pH of the cytoplasm away from the IEP of the chromosomes.

(1) *Meiosis*
(i) *Ammonia vapour* (PMC in 3% sucrose: 5–15 seconds before fixation).
(*a*) Aceto-carmine (Kuwada and Nakamura 1934).
(*b*) 2BE-crystal violet or Feulgen (La Cour 1935).
(ii) *Ammonia in 30% alcohol* (6 drops in 50 ml, 5–20 seconds). Flemming-crystal violet. Useful because alcohol attaches the cells to the slide (Sax and Humphrey 1934, Creighton 1938 on an amphibian, *Amblystoma*).
(iii) *Weak alkaline solvents of nucleic acid* (Oura 1936, Kuwada *et al*. 1938, Hillary 1940). Smear is placed in solutions of NaCN, NaHCO$_3$ or even 10^{-2} M NaOH for 15 seconds to 3 min according to the material. Flemming with CV or Feulgen.

(iv) *Acid fumes* (nitric, hydrochloric and acetic) followed by 2BE-CV (La Cour 1935).

(v) *Precipitation of desoxyribose-nucleic acid and partial digestion of protein:* Hillary's modification, 1940, of Caspersson's method.

(*a*) Pre-treatment as in (iii), 30 seconds in 10^{-2} M NaCN.

(*b*) Fixation in Flemming.

(*c*) Preparation left about 12 hours in 0·1% lanthanum acetate to precipitate nucleic acid as lanthanum salt.

(*d*) 24 hours at 37° C in 1% solution of trypsin containing a trace of lanthanum acetate.

(*e*) Stain by Feulgen.

(2) *Mitosis.* Most methods of pre-treatment are usually unsatisfactory for chromosomes within a cell wall. As a first attempt an ammonia–thorium nitrate method was suggested for root tips by Nebel (1934). Later Coleman (1940) was successful with treatment of smears of pre-meiotic mitosis using a solution of 10^{-5} M $NaCN_2$. Three other methods may sometimes overcome this difficulty:

(i) Pollen tube mitosis (Upcott 1936 on *Tulipa*). Aceto-carmine; penetration is easy on account of extremely thin cell wall.

(ii) Pollen grain, first mitosis (Geitler 1938, *cf.* Darlington and Upcott 1941). Pressure in fixation with aceto-carmine.

(iii) Root tip frozen, while growing, for four days to give differential reactivity of heterochromatic segments, which then show their spiral structure (see Ch. 9 *c*).

For revealing spiral structure in the chromosomes of cultured human leucocytes (Schedule 24), Ohnuki (1968) devised a special hypotonic solution consisting of equimolar (0·55 M) solutions of KCL, $NaNO_3$ and CH_3COONa in the proportion of 4:2:0·8, respectively. The cells, as a loose pellet, are re-suspended in the hypotonic solution and treated for 90 min at 21–23° C.

Finally micro-incineration (Barigozzi 1937, Uber 1940) has been used as a method of showing chromosome structure.

B. Micro-Chemical Tests

The effect of DN-ase on lampbrush chromosomes has shown that DNA runs continuously throughout the chromosome axis (Callan and Macgregor 1958). Histone is bound to DNA by salt-like linkages (Mirsky and Ris 1951) and is situated in the groove of the double-helix (Wilkins 1956). The fission of the chromosome axis with DN-ase indicates that other protein is attached laterally.

Proteins. A number of relatively simple histochemical methods for the demonstration of proteins in cells are available. Chromosomes of plant and animal cells have been shown to react more strongly than the cytoplasm but not differentially (Serra 1946). Recently the Sakaguchi reaction for arginine has been modified with the view that quantitative measurements might be made microphotometrically (Schedule 16, McLeish *et al*. 1957, McLeish and Sherratt 1958). Staining of tissues in fast green at pH 8 after removal of nucleic acids has been used for the estimation of basic protein in nuclei (Alfert and Geschwind 1953). It has recently been used by Bloch (1966) in combination with eosin to stain histones in animal tissues.

Nucleic Acids. Two types of nucleic acid are found in cells: (i) DNA, staining with Feulgen's reagent is present in the chromosomes, nucleolus, mitochondria and chloroplasts, and (ii) the RNA not staining with Feulgen consisting of three chemically distinct classes formed on DNA templates. The greatest concentrations occur in the cytoplasm particularly through movement from the nucleus as a prerequisite to protein synthesis. Some RNA appears to be present in the chromosomes throughout the whole mitotic cycle. The RNA present in the metaphase chromosomes may be derived largely from the nucleolus (Kaufmann 1948, La Cour 1963, Maio and Schildkraut 1967). The distribution of RNA in the cell largely governs, and is governed by, its activity and development, particularly in relation to protein synthesis (Caspersson 1941 and 1950, Brachet 1940*a*, 1940*b* and 1953, Painter 1943, Davidson and Waymouth 1944).

(i) *DNA*. Some measure of nucleic acid concentrations in cells can be obtained by staining methods. The Feulgen method is valuable for comparative evaluation of DNA content in nuclei within tissues and possibly between related species by microphotometry. Precautions must be taken to ensure the optimum reaction and to rigorously standardise staining conditions (*cf*. Kelley 1939, Stowell 1945, Stacey *et al*. 1946, Swift 1953). Recently McLeish and Sunderland (1961) have shown that the mean values obtained with Feulgen stain do not differ significantly from estimations of DNA per cell obtained biochemically. The quantitative aspects of microphotometry have been discussed by Swift (1953), Vendrely and Vendrely (1956), Swift and Rasch (1956), Leuchtenberger (1958). A photo-electric scanning device for Feulgen measurements, which is used in combination with a special cell crushing technique, has been described by Deeley (1955) (*cf*. Deeley *et al*. 1954).

The availability of actinomycin in tritiated form with high specific activity provides a sensitive means of detecting DNA autoradiographically in fixed cytological preparations in amounts too small to be shown by Feulgen staining (Camargo and Plaut 1967, Ebstein 1967). It is applied to sections, squashes or smears like a stain. It has been established that the drug binds selectively to guanine (see Reich and Goldberg 1964). Its binding to DNA can, therefore, give an indication of the distribution of guanine-rich DNA in the chromosomes (Avanzi and D'Amato 1972) and thus a means of detecting one of two main classes of heterochromatin (Ch. 9).

Formaldehyde or glutaraldehyde fixation should not be used, because the resulting cross linking of protein to DNA inhibits binding of the drug to the guanine.

(ii) *RNA*. A method devised by Brachet (1940a; *cf.* 1953) can be used for the detection and relative estimation of RNA. Its validity depends on the use of RNA–nuclease (R N-ase), in combination with Unna's methyl-green-pyronin stain. Cell constituents stain pink to red according to the concentration of RNA. The chromatin stains green or purplish according to the concentration of DNA and RNA. The nuclease-treated control should show no evidence of pyronin staining. The specificity of R N-ase has been examined by McDonald and Kaufmann (1954).

The activity of R N-ase is impaired by osmic and chromic acid types of fixation. Tissues should therefore be fixed in acetic alcohol or Carnoy. Air drying (see below) followed by hardening in 95% alcohol or methyl alcohol for 2 hours is sufficient for some types of cell, *e.g.* yeasts and blood. Squashes of embryo sac mother-cells and root tips can be subjected to Brachet's test if enzymatic methods are used for cell separation (Ch. 5 *d*.). Fixatives that preserve phospholipids in the cell seem to impair the differential qualities of Unna's stain (*cf.* La Cour *et al.* 1958, La Cour and Chayen unpub.). Improvements in the preparation of Unna's stain have been made by Taft (1951) and Kurnick (1955).

Azure B (Flax and Himes 1952; Schedule 13) at pH 4 stains RNA purple and DNA blue-green. With appropriate nuclease extraction of sections, smears or squashes, it provides a useful control in tests for the specificity of radioactive nucleic acid precursors in autoradiography. It has been used for relative estimation of RNA following removal of DNA with DN-ase (Woodard *et al.* 1961).

Phospholipids. Evidence for the presence of these substances in chromosomes has been obtained by histochemical and biochemical methods (La Cour *et al.* 1958, Chayen *et al.* 1958). (i) Tissues are fixed in Lewitsky's fluid and the sections stained by a modified acid haematin reaction. The euchromatic parts of chromosomes give a blue-black reaction for phospholipids from mid-prophase to end of anaphase in mitosis and meiosis, the heterochromatic regions give the same reaction at all times except early prophase. (ii) Yellow staining of phospholipid-containing regions is obtained when the sections are stained in a mixture of orange G-aniline blue at pH from 2–3. (iii) Methods for revealing the presence of phospholipids histochemically masked by protein interference have been suggested by Serra (1958), *cf.* La Cour *et al.* (1958) and Berenbaum (1958). The acid haematin procedure of Baker (1946) is most suitable for showing phospholipids in mitochondria.

For purposes of chemical tests of nucleic acids and proteins in animal cells, nuclei, chromosome threads, and such cytoplasmic particles as mitochondria can be separated with a high-speed centrifuge or by gentle homogenisation in a Waring blender (Claude 1941, Claude and Potter 1943, *cf.* Anderson 1956).

It is sometimes necessary to remove one or both nucleic acids from tissues for further histochemical procedures or for investigations in autoradiography. A choice of method is available:

(i) Digestion of RNA or DNA by specific enzymes (Schedules 14, 15).

(ii) Removal of both nucleic acids by treatment of sections or tissues in 4% trichloroacetic acid at 90° C for 15 min (Schneider 1945) or in 5% perchloric acid at 60° C for 30 min (Erickson *et al.* 1949).

(iii) Removal of RNA by treatment of sections or smears, after fixation in acetic alcohol or Carnoy, in N HCl at 60° C for 10 min or in 10% perchloric acid at 4° C for 18 hours (Erickson *et al.* 1949). The last method is not always satisfactory for plant material.

The theory and application of various histochemical tests are discussed by Pearse (1953).

With the help of these tests we can make a number of clear definitions. *Chromatin* is part of a chromosome thread to which the characteristic DNA is attached. A *chromosome* is a whole thread to which this acid is attached and having the characteristic capacities of spiralisation and replication in a regulated cycle. A *nucleolus* is a body within a nucleus consisting largely of gene product resulting from the activity of a specific loop-forming segment of a nucleolar chromosome (La Cour 1966), Plate XXIX. Free nucleoli, as in oocytes of amphibia, each contain a DNA replicate synthesised at the same locus as the attached nucleolus (Miller 1964, Callan 1966). *Heterochromatin* is a part of the chromosome which is out of step with the major bulk of *euchromatin* and replicates late in the cell cycle (Lima-de-Faria 1959, Keyl and Pelling 1963). Heterochromatin, unlike euchromatin, may remain tightly spiralised at telophase and throughout interphase (Heitz 1929, 1932) or as found in *Fritillaria lanceolata* remain spiralised at a constant intermediate level for the whole mitotic cycle (La Cour and Wells 1973); and also at meiosis from leptotene to zygotene (La Cour and Wells 1970). In animals, particularly mammals, heterochromatin is often associated with the centromere (*e.g.* Pardue and Gall 1970, Arrighi and Hsu 1971, Hsu and Arrighi 1971) or with the nucleolus as nucleolus associated chromatin (Caspersson 1950). The strong probability that heterochromatin exists in two main classes, with perhaps intermediate variants in respect of spiralisation, is favoured by recent studies on satellite DNA (see Yunis and Yasmineh 1971), inasmuch as these suggest that the DNA in non-nucleolar heterochromatin consists of repetitive base sequences, either AT or GC in constitution. Finally, with the aid of new techniques now available for recognition of heterochromatin (Schedules 18 and 19, Ch. 8 *m* and *n*), there should be less confusion with the phenomenon of negative and positive heteropycnosis.

C. Animal Eggs

Three stages in the development of the egg concern us:

(1) Prophase of meiosis: this begins in the young ovary and may last the greater part of its life.

(2) The meiotic divisions: these are rapid and are usually set in motion by the entry of the sperm.

(3) The cleavage division of the fertilised egg: these exceed in their rapidity all other mitoses and in consequence have a character of their own (Sonnenblick 1940).

During the oögonial divisions and in early prophase neither yolk nor shell is present to hinder treatment, and the ovary can be treated like a testis. Where the shell is thin and the yolk deficient, squash preparations can still be made in all the later stages. This is true of the Annelida and Mollusca. For example in *Allolobophora* eggs, Foot and Strobell (1905) obtained their best preparations of meiosis by puncturing each egg in a very small drop of water. When the contents of the egg were dragged out of its membrane they dried so rapidly as to give admirable fixation of the nucleus in a self-flattened smear. Twenty such eggs, *fixed by drying*, were laid on a slide ruled in squares with a diamond and stained with aqueous Bismarck brown. Alternatively they may be fixed and stained in acetic-orcein, or fixed in acetic alcohol and stained in Feulgen (Muldal 1952).

Where the shells are thick (as in grasshoppers) they may be removed by puncturing or by dipping for 2 min in 3% aqueous sodium hypochlorite (Slifer 1945). Otherwise the shells must be punctured if possible to admit the fixative and to aid the infiltration of wax. Alcoholic fixatives, particularly Kahle's fluid or S. G. Smith's (1941) modification, are the best. Special methods of dehydration and infiltration are necessary for obtaining good sections, owing to the brittleness of the yolk. Smith's (1941) procedure is perhaps the most favourable (Schedule 1″, *v*. Ch. 6 *e*). Haematoxylin or crystal violet (chromic method) staining are preferable for meiosis, Feulgen for cleavage divisions.

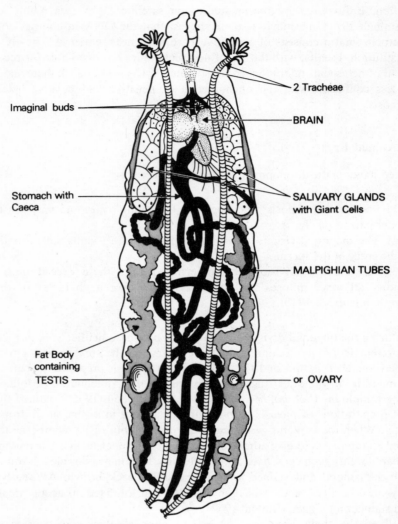

FIGURE 1 *Drosophila* **Larva**

D. Salivary Glands

Salivary glands and some other bland tissues of Diptera contain nuclei in a perpetual interphase. These were first described by Balbiani in 1880 but only understood fifty years later (Heitz and Bauer 1933, Painter 1934). Their nuclei are assumed to be polyploid on three grounds: their size, the multiple or polytene structure of their chromosomes, and their total DNA content which is a high multiple of that of ordinary mitotic nuclei. Their chromosomes, when fully extended, are about 100 times as long as at metaphase of mitosis. The 128, 256 or 512 threads of each homologous type are associated in parallel; each gene therefore appears as a band which is visible in the living state (Plate VIII). The fully developed banded chromosomes are as long as a quarter or half a millimetre and when stained are visible with a hand lens. Structural hybridity is shown by the same changes of partner as with pachytene pairing at meiosis. Salivary glands are important therefore both for study in combination with genetics and systematics and for micro-chemistry (*cf.* Caspersson 1940). Salivary glands can also be made use of in labelling experiments, such as with the use of radioactive precursors of DNA and RNA in demonstrating the metabolic activity of genes (Pavan and Ficq 1957, Pelling 1959, 1964).

Larvae of the most suitable species may be obtained as follows:

Chironomus (Poulson and Metz 1938) is the common blood-worm of ponds.

Bibio (Heitz and Bauer 1933) feeds on roots and leaf-mould.

Drosophila (Demerec and Kaufmann 1961), the fruit fly, may be trapped by a bait of over-ripe fruit, and the flies bred on a special pabulum (App. II *e*).

Sciara (Metz 1935), the dung fly, grows on fungi; can be bred on mouldy *Drosophila* pabulum.

Well-fed larvae ready for pupation have the largest chromosomes. Each salivary gland contains 28–32 cells in *Sciara*, 28–44 in *Chironomus*, 100–120 in *Drosophila*. These are in different stages of development—the largest in the middle or at the rear end (Fig. 1).

Dissection. Take the full-grown larva, having established its sex if necessary, and place it on a slide in a large drop of isotonic Ringer solution or stain fixative. Cut off its head with a needle in the right hand while pressing the body with a needle in the left. When the pressure is released the salivary glands will float out and can then be put on another slide (Schedule 4). Separate preparations can be made of ganglia, guts, or gonads as required (Koller 1940).

E. Tadpole Tails

The epidermal cells of the tail in the urodele larva (*e.g. Triturus*, the newt) are marked out for the study of mitosis. The cells are particularly favourable for prophase, since they are large and very flat. The best stage is when the tail fin is fully grown. Under natural conditions this stage is reached in

May, June and July. In the laboratory the larvae can be partially starved and brought into condition when they are needed.

The tail should be cut off just behind the anus and fixed in acetic alcohol for two to twenty-four hours. Preparations can later be made in iron aceto-carmine or acetic-orcein, or with Feulgen. Care should be exercised when using Feulgen not to press the preparation too hard lest the cells be crushed.

Larvae with amputated tails can be retained and used once more when the tail has regenerated. In Great Britain a vivisection licence is required for such operations.

F. Amphibian Testes

To study meiosis in the newt, testes should be fixed about a month after the breeding season. They should be cut into small pieces and fixed in acetic alcohol for two or three days. They are best stained with iron aceto-carmine. Very small pieces, the size being accommodated to that of the cover slip, should be placed on a slide in a small drop of stain, tapped out lightly, and covered after the material has darkened with the stain. The cover slip should be pressed lightly in the middle under blotting paper, and suitable isolated cells will be found lying at the edges of the preparation.

G. Lampbrush Chromosomes

Chromosomes of a type known as 'lampbrush' appear at a phase corresponding to diplotene in the ovarian oocyte nuclei of a wide variety of animals. Newts have the longest lampbrush chromosomes of all organisms and since these chromosomes are extended about twenty-five times the length of the chromosomes at metaphase of mitosis the longest lampbrush chromosome in *Triturus cristatus carnifex* is about 700–800 μm long (Plate XXIV). Lateral projections, consisting of pairs of loops some 20 μm or more when fully extended, are formed at chromomeres along the axis of these chromosomes. The growth and facility of loops to collect or discharge materials during growth of the oocyte reflects a metabolic activity of genes at the loci at which they form (*cf.* Callan and Lloyd 1956). The organisation and fine structure of these chromosomes has been studied extensively (Duryee 1937, 1950; Gall 1952, 1955; Guyénot and Danon 1953; Callan 1955, 1957; Wischnitzer 1957; Callan and Macgregor 1958).

Owing to their enormous length and fragile nature lampbrush chromosomes are best observed under phase contrast after isolation in saline. The germinal vesicle or nucleus from an ovarian egg is removed on to 0·1 molar KCl using forceps, needle and a binocular dissecting microscope. The nuclear membrane is freed from adhering yolk and cytoplasm by pipetting; the nucleus is then transferred to saline in a special observation chamber.

The observation chamber consists of a 3 in by 1 in slide with a $\frac{1}{4}$ in hole bored through the centre; a cover slip is placed across this hole and sealed to the slide with paraffin wax. For the newt *Triturus cristatus carnifex* a suitable saline is 7 parts 0·1 molar KCl : 3 parts 0·001 molar KH_2PO_4.

In the observation chamber the nuclear membrane is removed using a pair of the finest watchmaker's forceps and a tungsten needle pointed in molten $NaNO_2$. The nuclear sap disperses, the chromosomes become disentangled from one another and fall on the cover slip forming the bottom of the chamber. The preparation is covered with a second cover slip sealed in position with vaseline, to prevent evaporation of the saline, and examined with a phase contrast microscope having its optical train inverted. Since the chromosomes are in perpetual Brownian movement, photographs at high magnification must be taken with a flash.

H. *Culex* Testes

The pupa is placed on a slide in Ringer solution A and examined by dissecting microscope. Needles are placed in the head and tail, and the larva is broken in two as the needles are pulled apart. The testes with their small ducts attached, if the individual is male, are readily visible as small translucent egg-shaped bodies. They are removed with a fine needle to acetic alcohol, where they fix in two minutes. They should now be picked up by the needle and placed on a slide in a drop of iron aceto-carmine or aceticorcein. The preparations can be covered after about one minute's staining (the time should be adjusted to suit stain and material) and pressed lightly through blotting paper applied above the cover slip. Preparations can be made permanent. This method is suitable for most small insect testes.

An account of insect morphology may help in dissection, *e.g.* Imms 1938.

I. Blood

Red bone marrow is a useful source of mitosis in mammals. Moreover its study can provide some understanding of the causal relationship between mitosis and blood diseases (La Cour 1944). Active marrow is best obtained from the young of birds and small mammals by extraction from femoral bones. In man it is now obtained by sternal puncture.

New improved techniques are now available for handling these cells. In small mammals colchicine* can be injected intraperitoneally to accumulate metaphases and spread the chromosomes. This can then be followed by treatment of cells in hypotonic fluids which further aids spreading by expanding the cells (*cf.* Hsu 1952). Excellent preparations can be obtained

* Colcemid, a product of CIBA Laboratories Ltd, can be used instead of colchicine. Dr Ford has found it to be less toxic.

when this sequence of treatments is followed by fixation of the cells as a suspension (Ford and Hamerton 1956). A simple method involving direct colchicine treatment of aspirated marrow is described by Tjio and Whang (1962).

Peripheral blood is now the best source of mitosis in humans. It has been recently discovered that extracts from the seeds of *Phaseolus vulgaris* produce active cell multiplication after three days of incubation *in vitro* (Nowell 1960). The active agent is phytohemagglutinin.* Its mode of action is not yet properly understood. The blood cells which undergo mitosis are probably the monocytes and the lymphocytes. Blood is obtained by veni-puncture and excellent preparations of mitotic chromosomes can be obtained by processing it according to Schedule 24.

J. Mammalian Embryos

Methods for obtaining chromosome preparations from mammalian embryos have been described by Ford and Woollam (1963), Tarkowski (1966), Shaver and Carr (1967), Butcher and Fugo (1967), Wroblewska and Dyban (1969) and Evans *et al.* (1972). The last authors devised a rapid method for mouse embryos which utilises the membranes from implants of a gestational age of seven and a half or more days. The implants are dissected from the gravid uterus and treated with colcemid before the membranes are removed for subsequent hypotonic pre-treatment. After brief fixation in acetic alcohol, the membranes are transferred to 60% acetic acid, where the cells become detached. The free cells are then spread by drying the acetic acid on heated microscope slides. The spread cells are stained in toluidine blue-resin, a combined stain and mounting medium (Appendix II, Breckon and Evans 1969).

The method when modified is applicable to tail tips of 1–6-day-old mice.

K. Tumours

Modern techniques provide rapid and satisfactory means of studying mitotic activity in tumours for cancer diagnosis and therapy in man, mammals and birds. The simplest method is by acetic-lacmoid or orcein squashes (Koller 1942; Schedule 3A). More elaborate methods are necessary, however, for detailed chromosome study. Siliconed slides can be used in order to facilitate chromosome spreading in cells of ascites tumours (Levan and Hauschka 1952). Better results are obtained if the cells are pre-treated in hypotonic fluids (Sachs and Gallily 1955). The method suggested by Ford and Hamerton (1956) is suitable for all types of tumour cells where Feulgen staining is required.

* A product of Difco Laboratories, obtainable as a dried powder.

L. Diagnosis of Sex in Somatic Tissue

The cytological diagnosis of the sex chromosomes in resting somatic nuclei has made possible a new approach to the study of normal and pathological human sexual development. In animals and insects the sex chromosomes, when large, form distinct chromocentres which are often located at the nuclear membrane. Moore *et al.* (1953) found in biopsy specimens of human skin that female skin shows about 70% of nuclei with a single chromocentre (the X) and male skin about 5%. Many other human tissues have since been shown to present the same sex difference (see Danon and Sachs 1957, Hamerton 1961).

Squash, smear (Schedules 5 and 6) or section methods, according to practicability, may be used to prepare the tissues for study. Acetic alcohol (1 : 3), 95% alcohol or methyl alcohol are suitable fixatives for squashes or smears. A new fixative (Klinger and Ludwig 1957) consisting of 30 parts 95% alcohol, 20 parts formalin, 10 parts glacial acetic acid, 30 parts distilled water is suitable for tissues prepared as sections.

To avoid errors in diagnosis it is necessary to readily distinguish chromocentres from small nucleoli. Methods employing the Feulgen stain obviate such errors. Alternatively, the thionin method (Schedule 17) employed by Klinger and Ludwig (1957) may be used.

Pearson, Bobrow and Vosa (1970) have demonstrated that male nuclei can be positively identified at interphase by virtue of the fluorescence given by the Y chromosome when nuclei are stained with quinacrine dihydrochloride (see Schedule 19). In nuclei of buccal smears, single fluorescent bodies were observed in normal XY males and double fluorescent bodies in XYY males. However, such identification is limited to man and gorilla, since it is only in interphase nuclei of these species that brilliant quinacrine fluorescence of the Y chromosome is obtained.

The quinacrine-fluorescent spot in somatic cells of human males can also be seen in 50% of the sperm from normal individuals and has been claimed to represent the Y chromosome. Some observations concerning the findings in sperm are discussed by Evans (1972).

M. Giemsa Banding In Metaphase Chromosomes

A Giemsa staining method that preferentially stains centromeric heterochromatin in mouse chromosomes was first described by Pardue and Gall (1970). Its specificity is believed to depend on denaturation of the DNA in fixed air-dried preparations followed by annealing in warm saline prior to staining. The centromeric DNA in mouse is known to consist of highly repetitive base sequences (Waring and Britten 1966) and to anneal more rapidly than the non-repetitive DNA of the genome (Britten and Kohn 1968). This so-called C-banding technique has been variously modified to demonstrate both centromeric heterochromatin and other characteristic Giemsa (G-) banding in human (Arrighi and Hsu 1971, Sumner *et al.* 1971, Schnedl 1971, Sumner 1972) and other mammalian chromosomes (Hsu and Arrighi 1971). It can also be used on meiotic chromosomes (Sumner 1972, Gallagher, Hewitt and Gibson 1973).

The factors and mechanism involved in the induction of specific banding patterns in mammalian chromosomes have been examined by Kato and Moriwaki (1972), Comings, Avelino, Okado and Wyandt (1973), Sumner, Evans and Buckland (1973) and Sumner and Evans (1973). Although it is clear that G-banding does not have to be associated with DNA made up of repetitive base sequences (Cullis and Schweizer 1974), the Giemsa stain is seemingly bound exclusively to DNA. The reason for the disparate behaviour along the chromosomes remains obscure.

An exact correspondence in G-banding pattern has been produced in human chromosomes by the use of trypsin and other proteolytic enzymes prior to Giemsa (Wang and Fedoroff 1972, Dutrillaux 1973) or Leishmann staining (Seabright 1972). See Schedule 18A.

Similarly, specific banding is inducible in human chromosomes by means of potassium permanganate prior to Giemsa staining (Utakoji 1972).

Finally, Dutrillaux (1973) has shown that preferential staining of the characteristic bands can be reversed (R-banding) in mammalian chromosomes by treatment of the chromosomes in spread preparations with Earles' medium at pH 6·5 for 10–20 minutes at 87° C prior to Giemsa staining. In other words, with R-banding, it is the remaining bulk of the chromosomes which becomes intensely stained.

G-banding was at first more difficult to apply to plant chromosomes inasmuch as conventional N HCl hydrolysis, so valuable in obtaining flattened meristematic cells in squash preparations, inhibits selective Giemsa staining. For the same reason cell separation in pectinase or snail cytase is unsuitable. Good results have been obtained, however, in squash preparations of root tips by using 40 or 45% acetic acid for maceration (Vosa and Marchi 1972, Schweizer 1973, Marks and Schweizer 1973) or following hydrolysis for 25–30 seconds in N HCl at 60° C (Vosa 1973). More consistent results are obtained when fixed material is steeped in 45% acetic acid for 15–30 minutes at 60° C prior to maceration (Marks unpub.).

Excellent preferential staining of bands is obtained in squash preparations of plant chromosomes when the DNA is denatured in a saturated solution of barium hydroxide for 5–8 minutes at room temperature (Vosa 1973, Marks and Schweizer 1973) prior to annealing in warm saline (Schedule 18). Schweizer found an exact correspondence in the position of Giemsa bands and H-segments previously noted in the chromosomes of certain plant species *e.g. Trillium grandiflorum* (Darlington and La Cour 1940) and *Fritillaria lanceolata* (La Cour 1951). Giemsa banding is evident also at prophase in both plants and animals, and chromocentres are preferentially stained at telophase and interphase.

N. Recognition of Heterochromatin with Fluorochromes

The use of fluorochromes in the study of the linear organisation in chromosomes was first described by Caspersson, Foley *et al.* (1968) and Caspersson, Zech *et al.* (1969 *a* and *b*). They observed that intense fluorescence occurred regularly at specific loci in the chromosomes of certain plants, when preparations were stained with acridine derivatives and especially quinacrine and its mustard. The position of these loci in *Trillium* chromosomes was found to more or less correspond with those of H-segments described by Darlington and La Cour (1940). Vosa (1970, 1971) has since studied the fluorescence given by various fluorochromes in the heterochromatin of several plants. He classified four classes of heterochromatin, according to whether they gave 'enhanced' or 'reduced' fluorescence with quinacrine and responded negatively or positively to cold treatment.

The phenomenon of enhanced and reduced fluorescence in H-segments is as yet not fully understood. Two investigations have suggested that the fluorescence of quinacrine, in the presence of DNA, is intensified where there is relatively high AT content and quenched where the DNA is correspondingly rich in GC (Weisblum and de Haseth 1972, Pachmann and Rigler 1972). The two categories of fluorescence may, however, reflect differences in states of chromatin condensation peculiar to two main classes of heterochromatin (La Cour and Wells 1973). There is clearly, though, a close correspondence between Giemsa banding and the segments distinguishable by quinacrine fluorescence, irrespective of category (Vosa and Marchi 1972).

The fluorochromes, as 0·5–1% aqueous solutions, can be applied to squash preparations of testes, salivary glands and neural ganglia of Diptera, suspensions of mammalian cells prepared as films, as well as squash preparations of RT macerated and squashed in 40% acetic acid (Schedule 19). The stained cells are examined in temporary preparations mounted in distilled water. A microscope fitted for fluorescence microscopy, with a mercury lamp, excitor and barrier filters is required (see Ch. 2).

In squash preparations, particularly of plant material, incident illumination is usually best because background debris is then invisible.

O. Identification of Sister Chromatids by Differential Staining

Methods have recently been developed which allow the identification of the two chromatids in chromosomes by differential staining and consequently the observation of sister chromatid exchanges. This development stems from the observations of Zakharov *et al.* (1974) showing that the sister chromatids of chromosomes in mammalian cells become disparately condensed at metaphase, after two rounds of replication in the presence of 5-bromodeoxyuridine (see *c. Differential Reactivity* Ch. 9).

The disparity, evidently a legacy of bifilar and unifilar substitution in the DNA, is further recognisable by the differential staining behaviour that is obtained with Giemsa (Ikushima and Wolff 1974) and the fluorochrome Hoechst 33258 (Latt 1973). The Giemsa and Hoechst dye, respectively stain and fluoresce weakly in the under-condensed chromatid arising after bifilar substitution, and conversely so in the fully condensed and unifilarly substituted sister. The result is a distinct harlequin effect (Plate XXIII).

The fluorescence produced by the Hoechst dye is unfortunately unstable and fading is rapid in light. Differential staining of the sister chromatids can, however, also be observed after staining with acridine orange (Perry and Wolff 1974), but the difference between chromatids is not readily visible until the cells are exposed to exciting light passed through a BG12 filter. Differential fluorescence is not obtainable with the fluorochrome quinacrine (Latt 1973).

A technique perfected by Perry and Wolff (1974), in which the use of the Hoechst dye is combined with Giemsa staining, to make permanent preparations is given in Schedule 27.

P. *In Situ* Localisation and Characterisation of Different Classes of Chromosomal DNA

It has been found that chromosomal regions containing satellite DNA stain distinctively from other parts of the chromosomes, after staining with a 0·5% solution of acridine orange in McIlvaine buffer (pH 7), subsequent to denaturation of the DNA with alkali (Chapelle, Schröder and Selander 1973). The distribution of double-stranded and single-stranded DNA in the chromosomes *in situ* can be assessed from the colour of the fluorescence of acridine orange in light with a wavelength of approximately 510 nm (Rigler 1966, Chapelle *et al.* 1971). Double-stranded DNA fluoresces green and single-stranded DNA red.

Denaturation of the DNA with alkali in metaphase chromosomes results in a shift in colour of fluorescence from green to yellow, orange, brown and red, as denaturation progresses.

In chromosomes of mouse and the vole, *Microtus agrestis*, the DNA denatures uniformly with alkali and in chromosomes of human cells sequentially (Chapelle *et al.* 1973). Further, after prolonged re-association in SSC saline, areas containing repetitious DNA, as in mouse centromeres, fluoresce more intensely with acridine orange than other parts of the chromosomes.

Q. Embryo Sac and Endosperm

Meiosis in the embryo sac may take place as long as two or three weeks after meiosis in the anther. In *Fritillaria* and *Tulipa* it usually coincides with the pollen grain division (*cf.* Darlington and La Cour 1942).

Wherever possible the ovules should be removed from the ovary before fixation. In a number of Monocotyledons they come out in entire strings. Very large ovaries should have part of the ovary wall removed.

The squash method after acetic alcohol fixation, either with Feulgen or acetic-lacmoid staining, is most suitable where cells and chromosomes are large (La Cour 1947). In favourable plants it is possible to tap out the cells, *e.g.* in 45% acetic acid after Feulgen staining, and remove unwanted tissue with the aid of a dissecting microscope before squashing. For small chromosomes and material difficult to handle the paraffin method has to be used, with a strong and penetrating aqueous fixative, *e.g.* 2BX.

Endosperm divisions are to be found a few days after pollination. In some species, *e.g. Scilla sibirica* and *Lilium* the divisions are synchronised for about six mitotic cycles and later with wall formation partially synchronised in patches.

Endosperm tissues are best handled as Feulgen squashes, the tissue being dissected out entire after staining (Rutishauser and Hunziker 1950; La Cour 1954; Schedule 5A). Dissection is performed under a dissecting microscope, using finely-pointed tungsten needles. The stage of mitosis and intensity of stain can be determined during this operation. Weak staining can be corrected by applying acetic-orcein to the dissected tissue. Longer hydrolysis times than those required for other tissues are necessary for some species and, in any case, after long storage.

In Coniferales it is possible to dissect entire endosperms from female cones before fixation (Sax and Sax 1933). At the right stage they look like the pupae of ants and reach this stage in northern countries during May. Fix in acetic alcohol and stain as squashes with acetic-orcein (Schedule 3).

R. Pre-Meiotic Mitosis

Pre-meiotic mitoses are to be found in the earliest stages of anther and embryo-sac development. In the anthers the mitoses are synchronised at least locally. Fix dissected organs, where possible, in acetic alcohol and treat either as Feulgen squashes or stain as squashes with acetic stains (Schedules 3 and 5).

S. Pachytene

The pachytene stage in pollen mother cells has been used for exact comparison of chromomeres, centromeres, nucleolar organisers and heterochromatin in many flowering plants. Special techniques have been developed. The following are representative examples:

Centromeres	*Agapanthus*	Darlington 1933a
Centromeres (diffuse)	*Luzula*	S. W. Brown 1954
Chromomeres	*Secale*	Lima-de-Faria 1952
Nucleolar Organiser	*Zea mays*	McClintock 1934
Heterochromatin	*Cicer*	Thomas *et al.* 1946
Heterochromatin	*Lycopersicum*	S. W. Brown 1949
Pairing: 2x	*Zea mays*	McClintock 1931, 1933
Pairing: 4x	*Hyacinthus*	Darlington 1929
Pairing: 4x	*Tulipa*	Upcott 1939
Pairing: 4x	*Lycopersicum*	Gottschalk 1955

9

The Control of Mitosis

A. Irradiation

Mitosis is controllable by irradiation and by chemical agencies (including nutritional deficiencies) and by the combination of the two. The chemical agencies are often highly specific. The irradiation effects act always on the whole cell—both directly and indirectly.

(1) X-rays

The specific effect of X-rays on the cell in producing gene mutations and chromosome breakages was first shown by Muller in 1927 and by Stadler in 1928. It at once became a most important experimental method of handling, or rather mishandling, chromosomes.

The apparatus required for studying qualitative effects of X-rays is simple. For quantitative tests of effects of varying dosages, intensities, and wavelengths, more elaborate equipment is necessary.

The minimum equipment consists of an X-ray tube together with the necessary rectifier, transformers and controls. The desired dosages can be obtained by trial and error, which indeed must be the basis of attack on any new material, for the effective dosage varies enormously between the highly susceptible root-tips of *Trillium* whose chromosomes give the most favourable breakage frequencies at 45 R, resting seeds which may resist 10 000 R without serious damage (*cf.* Gustafsson 1937), and the multinucleate conidia of *Neurospora* only half of which are killed by 500 000 R. Polyploids also show much less effect from mutation or chromosome breakage than do related diploids with the same dose (*cf.* Müntzing 1941).

A characteristic intensity shown by a Coolidge water-cooled tube is as follows:

75 R min^{-1} at 30 cm target distance, with 72 kV means e.m.f. and 5 mA current.

The material, methods, dosages and stages used for treatment vary according to its purpose (*cf.* Stadler 1930, Knapp 1935, Gustafsson 1936, Goodspeed and Uber 1939) as shown in Table III. But for the immediate study of mitosis with chromatid breakage we can recommend lower doses than have commonly been used (50–100 R).

The immediate effects of X-rays depend on the stage of development treated. Resting seeds are delayed in germination. Nuclei in prophase of mitosis are set back into resting stage, to return later in a polyploid condition. Sperm nuclei of animals or plants can reveal no immediate effect. The delayed effects that are observed depend on the length of time after treatment during which the changed cells and changed chromosomes have to survive if they are to be observed. The more serious changes, especially in haploid pollen, kill the cell after mitosis has led to the loss of fragments without centromeres and the breakage of chromosomes with two centromeres (*cf.* Darlington 1937, Muller 1940, Darlington and Upcott 1941).

The efficacy of X-rays in producing aberrations in chromosomes can be modified by the oxygen tension in tissues during irradiation. This was first shown by Thoday and Read (1947) in roots of *Vicia*, and their conclusion that the frequency of aberrations is reduced at low oxygen tension has since been confirmed in various experiments involving anoxia (Hayden and Smith 1949, Giles and Riley 1950, Giles *et al.* 1952, Swanson and Schwartz 1953 and others). Recently, treatment with chemical reducing agents before irradiation has been shown to produce similar results (Wolff 1953).

It will be seen that X-ray studies, as is proper in their initial stages, have been directed to organisms with large chromosomes capable of showing precise effects. There is another group of observations however which stand apart in this respect—X-ray work on human cancer. In this field the introduction of smear methods of studying mitosis has greatly improved the diagnosis of malignancy in tumours and also the accuracy with which curative doses of X-ray and radium treatment can be applied (Ch. 8).

(2) *Other types of ionising radiation*
(i) Gamma-rays like X-rays are electromagnetic radiations which produce charged particles in the cell. They are obtained either from radium or from the radioactive isotope Co^{60}. Gamma-rays are more penetrating than X-rays, and since Co^{60} provides a continuous source of radiation it is being utilised for producing mutations in plants by growing them under conditions of continuous exposure (see Sparrow and Christensen 1953, Sparrow and Singleton 1953).

(ii) Beta-rays can either be obtained from special generators or from certain radioactive isotopes, *e.g.* P^{32}, C^{14}, etc. The radiations emitted consist of charged particles having low penetration which somewhat restricts their application in mutation and chromosome breakage studies.

TABLE III – BREAKAGE OF CHROMOSOMES BY X-RAYS

Organism	Organ and Cell	Stage		Dose	Type of Breakage, etc.	Reference
		Treated	Observed			
ANIMAL						
Drosophila	adult fly	sperm	offspring	1000 R–5000 R	general	Muller '40
Drosophila	adult fly	sperm	offspring SG	1000 R–5000 R	H-segments	Kaufmann '46
Drosophila	egg (screened)	nucleus	cleavage	500 R	lethality	Ulrich '57
Drosophila	egg (screened)	cytoplasm	cleavage	105 R	lethality	Ulrich '57
Drosophila	embryo	SG	adult	5000 R	—	Geyer-Duszynzka '55
Locusta	testis, adult	mitosis RS	met: X_2	ca. 500 R	diplochromosomes	White '35
Chortophaga	embryo	mitosis RS	living mitosis	4–8000 R	general	Carlson et al. '53, '55
PLANT						
Trillium	RT	RS	met	5–45 R	B", no R"	D. and L.C. '45
Hyacinthus	RT	RS	met: X_2	150–1000 R	B" + R"	La Cour '53
Vicia	RT	RS	met	50–200 R	B", no R"	D. and L.C. '45
Vicia	RT	RS	met		sub-chromatid	Thoday '53
Vicia	RT	prophase	met	100 R	sub-chromatid	Davidson '57
Vicia	RT	metaphase	met: X_2	150 R	B" + R"	Davidson '58
Tradescantia	anther	before meiosis	PG: met	150 R	cell selection	Sax '38; cf. D. '52
Lilium	PMC	meiosis, pro. I	Met I, II.	150 R	B" + R"	Sauerland '56
Uvularia	PMC	meiosis, pro. I	met-ana I	90 R	sub-chromatid	D. and L.C. '53
Tradescantia	PMC	meiosis later	met II.	18 R	all types	Haque '53
Tradescantia	PMC	meiosis later	PG: met	360 R	B" + R"	Haque '53
Trillium	PMC	meiosis later	PG: met	50 R	DNA estimation	Sparrow et al. '52
Trillium	PG early	RS	PG: met	45–375 R	B", rare R"	D. and L.C. '45
Tradescantia	PG early	RS	PG: met	360 R	B" + R"	D. and L.C. '45
Tradescantia	PG early	RS	PG: met	30–300 R	(split doses)	Lane '51, Sax et al. '55
Tradescantia	PG early	prophase	PG: met	25 R, 3° C	sub-chromatid	Sax and King '55
Tradescantia	PG late	RS	PT met	180 R, 250 R	B", no R"	Catcheside and Lea '43, D. and L.C. '45, Bishop '55
Zea mays	PG ripe	sperm	meiosis, F_1	800–1500 R	interchanges	Catcheside '38b
Scilla	endosperm	prophase	ana	50 R	sub-chromatid	La Cour and Rutishauser '54

Note.—X_2 is the second mitosis after treatment with X-rays, etc.

TABLE IV – BREAKAGE OF CHROMOSOMES BY OTHER RADIATIONS

Organism and Part	Cells affected	Time of Observation	Treatment and Dosage	Reference
GAMMA RAYS				
Tradescantia (2x)	PG	PG mit	5 R min^{-1}; 100–2000 min	Koller '53
Tradescantia (2x)	PG	PT mit	Co^{60}, 100–400 R	Kirby-Smith and Daniels '53
Tradescantia (2x)	PG	PG mit	Co^{60} 1·1–1·3 MeV (in air and nitrogen)	Swanson '54
BETA RAYS				
Tradescantia (2x)	PG	PT mit	External P^{32} plaque 102–38 rep	Kirby-Smith and Daniels '53
Tradescantia (2x)	anthers	PG mit	Internal, absorption through stem (i) C^{14} from $(NH_4)_2 CO_3$ 0·9–8·2 μc/ml 4–8 days (ii) P^{32} from $Na_2(HPO_4)$ 1–10 μCi/ml 24 hr–9 days	Giles and Bolomey '48
Drosophila	whole larva	progeny	P^{32}* in food *ca.* 1 mg per fly	Bateman and Sinclair '50
ALPHA RAYS				
Tradescantia (2x)	PT	PT mit	various, from polonium	Catcheside and Lea '43
Tradescantia (2x)	PG	PG mit	stripped inflorescence in radon soln.	Kotval and Gray '47
Vicia faba	RT	up to 3 days, mit	in radon soln. 7–8 energy units	Thoday and Read '47

NEUTRONS				
Tradescantia (2x)	PG	PG mit	various	Thoday '42
Tradescantia (2x)	PG	PG mit	various	Giles '43
Drosophila (2n = 8) ♂	whole fly	progeny	various	Giles '43
ULTRAVIOLET				
Tradescantia (2x)	PT	PT mit	wavelength 2540 Å 2×10^{-3} erg mm^{-2}/60s	Swanson '43
Tradescantia (2x)	PT	PT mit	before and after X-rays, wavelength 2537 Å	Swanson '44
Zea mays (2x)	PG: sperm nuclei	Endosperm phenotypes	wavelength 2537 Å 546 000 erg cm^{-2}	Fabergé '56
NEAR INFRARED				
Tradescantia (2x)	PG	PG mit	10 000 Å *ca.* 3 hr before and after X-rays 107 R	Swanson '49
Tradescantia (2x)	PG	PG mit	10 000 Å *ca.* 3 hr before and after X-rays 90 and 350 R	Yost '51
Tradescantia (2x)	PMC: zygotene -pachytene	meiosis: MI	10 000 Å *ca.* 3 hr	Snoad '88
Drosophila (2n = 8) ♂	sperm	SG progeny	10 000 Å *ca.* 72, 144 and 216 hr (2000 R X-rays—Infrared—2000 R X-rays)	Kaufmann *et al.* '46
Drosophila (2n = 8) ♂	sperm	SG progeny	10 000 Å *ca.* 48 hr before and after X-rays 3000 R	Kaufmann and Gay '47

* See Williams and Dowrick (1958) in relation to uptake, distribution and mutation in plants.

TABLE V – CHEMICAL AND OTHER NON-RADIATION BREAKAGE OF CHROMOSOMES

	Treatment		Organism and Part	Time of Observation	Reference
	Concentration	Duration			
BREAKAGE RANDOM					
Nitrogen mustard	vapour, pure or dilute	2–3 min	*Tradescantia*, anthers	Meiosis, PG: met-ana (centromere)	D. and Koller '47
Acridines	0·1–0·01 %	4 hr–3 days	*Allium*, RT	met-ana	D'Amato '52
Purine derivatives	0·5–50 mM	4–24 hr	*Allium*, *Pisum*, RT	met-ana	Kihlman '52
Oxygen	50–100% at various pressures	3 hr on flowers 1 hr on PG	*Tradescantia*, anthers	Meiosis, PG and PT mitosis	Conger and Fairchild '52
Ethylurethane	0·1 M + 0·05 M KCl	1–48 hr	*Zea mays*, ears	Pachytene	Linnert '50
Coumarin	1mM and with colchicine	12 hr at 13°C	*Allium*, RT	met-ana	Östergren and Wakonig '54
Coumarin and derivatives	0·5–2 mM	4 hr–3 days	*Allium*, RT	met-ana	D'Amato *et al.* '54
Ethylene oxide	0·6%		*Hordeum*, seeds (wet and dry)	progeny	Ehrenberg *et al.* '57
Diepoxybutane	0·003%	2–10 hr			
BREAKAGE LOCALISED **(i) *In or near hetero-chromatin***					
Maleic hydrazide	0·1–10 mM	1–20 hr	*Vicia*, RT	met-ana	McLeish '53, '54
Di(2:3-epoxypropyl) ether	up to 10^{-3} M	10 min–10 hr	*Vicia*, RT	met-ana	Revell '55
Potassium cyanide	5×10^{-4} M	30 min in N_2 and O_2	*Vicia*, RT	met-ana	Lilly and Thoday '56
Potassium cyanide	10^{-3}–10^{-4} M	½–5 hr in air and various O_2 concs.	*Vicia*, RT	met-ana	Kihlman '57
Low temperature	0°C	3 days	*Trillium*, RT	anaphase (sub-chromatid)	Shaw '58

(ii) *In or near nucleolar organiser*					
8-ethoxycaffeine (EOC)	5–7·5 mM		*Vicia, Pisum*, RT	met-ana	Kihlman and Levan '51
8-ethoxycaffeine (EOC)	10^{-2} M and with sodium azide	$\frac{1}{2}$–8 hr at 10° and 20°C in air and N_2	*Vicia, Pisum*, RT	met-ana	Kihlman '55
(iii) *At ends*					
Storage (ageing)	—	3 weeks	*Kniphofia*, pollen	PT : ana (SR)	Barber '38
(iv) *In regions stable to X-rays*					
Alkylating agents imines, epoxides	10^{-2} M:0·3–0·4 µl per fly	—	*Drosophila* ♂ 25–35 hr old	progeny and SG	Fahmy and Bird '56

NOTES ON TABLE V

1. Nitrogen mustard is NN-di(2-chloro-ethyl) methylamine.
2. Maleic hydrazide is 2 : 6 pyridazine diol, an isomer of uracil, a pyrimidine base of RNA.
3. SPONTANEOUS BREAKAGE, due usually to unspecified genotype-environment reactions; has been described in plants in what may be regarded as *terminal* cell divisions.

Recent references are:

(i) MEIOSIS: *Tradescantia*, Giles '40; *Tulipa*, etc., D. and Upcott '41. Usually associated with asynapsis; not to be confused with regarded as *terminal* cell divisions.

(ii) PG MITOSIS: *Scilla*, Rees '52; *Allium*, D. and Haque '55; *Paeonia*, J. L. Walters '56; *Bromus*, M. S. Walters '57.

(iii) ENDOSPERM: *Trillium*, etc., Rutishauser and La Cour '56; *Zea mays*, Schwartz '58.

TABLE VI — CONTROL OF MITOSIS BY CHEMICAL AGENTS

Reagent	Concentration	Duration	Part	Plant	Reference
A. Production of polyploids					
Colchicine	0.1% in 0.8% agar	3 days	buds	*Datura*	Blakeslee and Avery '37
Colchicine	0.2–1.6% aqu.	12 hr–10 days	seeds	*Datura*	Thomas '45
Colchicine	1% in lanolin	smeared	buds	*Antirrhinum*	Nebel and Ruttle '39
Colchicine	0.01–0.2% aqu.	17 hr	runners	*Mentha*	Nebel and Ruttle '39
Colchicine	1% in 1.5% agar*	smeared	buds	*Petunia*	Levan '39
Colchicine	0.125% aqu.	20 hr†	seeds	*Petunia*	Levan '39
Colchicine	0.06% aqu.	2 days	germinating seeds	*Hordeum*	Karpechenko '40
Colchicine	0.05–0.1% aqu.	12–24 hr	dry seeds‡	*Gossypium*	Harland '40
Colchicine	0.05–0.1% aqu.	12 hr	shoots	*Gossypium*	Harland '40
Colchicine	0.05% aqu.	48 hr, 28°C	dissected embryos	*Pyrus malus*	Thomas '45
Acenaphthene	Satd. soln. on paper	2–7 days	shoots	*Antirrhinum*	Kostoff '38
Nitrous oxide (Laughing gas)	10 atm pressure in iron vessel	4–6 hr	whole plant (1st division of embryo)	*Crepis*	Östergren '54
B. Examination of mitosis					
Aurantia	0.01–0.1% aqu.	1–6 days	roots	*Hordeum*	Favorsky '40
Tribromoanaline	powder	sprayed	roots	*Hordeum*	Favorsky '40
Colchicine	0.125–2% aqu.	7 min–72 hr	roots	*Allium*	Levan '38
Colchicine	0.1% aqu.	to 2 days	larva	*Triton*	Barber and Callan '42
Naphthalene derivatives	aqu. solns. various conc.	12–18 hr	roo.s	*Allium*	Levan and Östergren '43
Para-dichlorobenzene	sat. aqu. soln.	1–4 hr	roots	various spp.	Meyer '45

Monobromobenzene	sat. aqu. soln.	2–4 hr	roots	various spp.	O'Mara '48
α-monobromo-naphthalene	sat. aqu. soln.	2–4 hr	roots	various spp.	O'Mara '48
8-hydroxyquinoline	0·002 M aqu. soln.	3–6 hr at 18°C	roots	various spp.	Tjio and Levan '50
'Gammexane'	sat. aqu. soln.	2 hr	roots	various spp.	D'Amato '52
'Coumarin'	2% aqu. soln.	2 hr at 12–16°C	roots	various spp.	Sharma and Bal '53
C. Activation of mitosis by hormones					
Heterauxin	1–10 p.p.m. aqu.	4–6 days	roots	*Allium*	Levan '39d
Naphthalene-acetic acid	0·25% in lanolin	10–20 days	stem inter-nodes	*Phaseolus*	Dermen '41
Naphthalene-acetic acid	0·000–0·0002–186·2 p.p.m. aqu.	14–22 days	roots	*Allium*	Levan and Lotfy '49
Indole-3-acetic acid	20–50 p.p.m. aqu.	2–4 hr	roots	*Rhoeo*	Huskins and Steinitz '48
Indole-3-acetic acid	3 p.p.m. 30 p.p.m.	1–7 days 4–8 hr	roots	*Allium*	Therman '51
Sodium 2-4-dichloro-phenoxyacetate	10–500 p.p.m. aqu.	3–5 days	roots	various spp.	Avanzi '51; D'Amato '52
D. Partial synchronisation of mitosis					
5-Aminouracil	700 p.p.m. half-strength Hoaglands nutrient soln.	24 hr§	roots	*Vicia*	Smith, Fussell and Kugelmann, '63
E. Control of Meiosis					
Colchicine	0·5% aqu.	from 7 days	flower buds	*Fritillaria*	Barber '42

* Warm.   † Vacuum-pumped.   ‡ Germinated afterwards at 30–35° C (Harland unpub.).

§ Seedlings then washed for 15 min and roots fixed 14 hr after recovery, or after 7 hr treatment in 0·01% colchicine (Trosko and Wolff, 1965).

(iii) Alpha-rays are given off by a few radioactive substances, notably radon and polonium. The radiation emitted consists of charged particles of high ion density and low penetration.

(iv) Fast neutrons do not produce ionisation directly but knock out protons from the nucleus of the atom they traverse. They are usually obtained from a cyclotron or atomic piles, or indirectly from special generators.

(3) Non-ionising radiations

(i) Ultraviolet radiations for biological experiments are obtained from a mercury lamp or from a quartz mercury arc in conjunction with a mono-chromator. Because of their long wave length (3900 Å to 1800 Å) their penetration is restricted. Their biological action is seemingly due to energy being absorbed by specific substances in the cell, particularly the nucleic acids. Ultraviolet produces gross physiological disturbances in cells.

(ii) Infrared radiations for biological experiments can be obtained from special types of tungsten lamps used for rapid drying purposes in industry. Suitable filters are interposed to remove the unwanted visible wavelengths. Near infrared (of wavelengths about 10 000 Å) has been found to modify the frequency of aberrations produced by X-rays. Its action, most likely, is to favour new reunion over restitution, possibly by influencing chromosome movement.

Some references to the kind of experiments performed with these types of irradiation are given in Table IV. A bibliography on the effects of ionising radiations on plants has been compiled by Sparrow et al. (1958). The subject of chromosome aberrations induced by ionising radiations has been exten-sively reviewed by Evans (1962).

The consequences of irradiation for the cell interact on one another. Special precautions in experimental design are therefore needed but are frequently neglected. The following should be noted.

Since breakability varies enormously during the mitotic cycle, it is of the first importance in breakage studies that times should be accurately recorded, temperatures accurately controlled and results completely described (Darling-ton and La Cour 1945). Further, all breakages arise from the same treatment whether they are followed by reunion or not. All the breakages must there-fore be recorded and recorded separately from the reunions in which some of them are involved. The problem of estimating delayed breakage from metaphase irradiation has been broached by Davidson (1958).

B. Chemical Agencies

(1) Chromosome breakage

The mutagenic effects of chemical agents were first discovered genetically with mustard gas in Drosophila (Auerbach and Robson 1946). Mustard gas was also shown to cause breakage of chromosomes when applied to anthers of *Tradescantia*. In the resting stage the effect resembles breakage produced by ionising radiation. At metaphase it induces misdivision of the centromere (Darlington and Koller 1947) as does low intensity radiation (Koller 1953).

Since these early observations numerous substances have been found to produce similar results (see Oehlkers 1952, Levan 1951, D'Amato 1952, von Rosen 1955). Deficiencies either in Ca- or Mg-ions, which in chromosomes are thought by Mazia (1954) to link the macromolecular complexes of nucleic acid and protein together, have also been found to produce chromosome fragmentation (Steffenson 1953, 1955). It may be noted, however, that chromosome breakage does not occur when roots are grown in chelating agents which should remove these metallic ions (Davidson 1958, *cf.* McDonald and Kaufmann 1957).

From these studies we know that some chemical agents produce chromosome breakage but not mutations and *vice versa*. Also that some break the chromosomes at specific regions and others at random. Some of these agents are listed in Table V.

(2) *Sister chromatid exchanges*
Evidence of sister chromatid exchanges in autoradiographs of tritium-labelled chromosomes was first reported by Taylor (1958), and since then evidence of such exchanges has similarly been obtained many times in both plant and animal cells (*e.g.* Olivieri and Brewen 1966, Geard and Peacock 1969 and Rommelaire, Susskind and Errera 1973). It is generally agreed that a high proportion and sometimes all of such exchanges may be induced by beta radiation from the tritium required for their autoradiographic detection. It is also extremely likely that they occur together with chromosome breakage produced by other means.

A means of observing sister chromatid exchanges without the use of autoradiography is now made possible with the introduction of methods which allow discrimination of sister chromatids by differential staining (Latt 1973, Perry and Wolff 1974, Schedule 27). This disparate staining behaviour arises from unequal condensation of sister chromatids, following incorporation of 5-bromodeoxyuridine (BUdR) or 5-iododeoxyuridine (IUdR) into the DNA of chromosomes during two rounds of replication (Zakharov *et al.* 1974, Ikushima and Wolff 1974, Perry and Wolff 1974, see also Ch. 8).

So far, these experiments have been confined to mammalian cells. The analogue BUdR is known to produce chromosome breakage in such cells (Hsu and Somers 1961, Dewey and Humphrey 1965), so that in addition to a low frequency of sister chromatid exchanges which may arise spontaneously, some may be BUdR-induced.

It seems probable that the frequency of such induced exchanges would be influenced by how much the incubating cells are exposed to light. It is known that BUdR sensitises DNA to light of wavelengths above 300 nm. Regan, Setlow and Ley (1971), and Ikushima and Wolff (1974) have shown that, in treated Chinese hamster ovary cells, the number of exchanges is dramatically increased when cells are exposed to light from an electronic flash unit, discharged at two-second intervals during incubation.

Finally, it is clear that some sister chromatid exchanges must arise spontaneously, as indicated by studies with ring chromosomes that form dicentric rings after an odd number of exchanges (Brewen and Peacock 1968).

(3) *General*

Chemical agents are being used with success as spindle anaesthetics to suppress the division of nucleus and cell. The ulterior objects are (i) to produce polyploid tissues and organisms and (ii) to examine their mode of origin in the cell and show how the spindle works, (iii) to suppress or modify chromosome pairing at meiosis, and (iv) to arrest mitosis at metaphase and thereby facilitate the study of metaphase chromosomes.

For producing polypoid plants treatment of dormant or relatively inactive tissue is least effective. Effectiveness is directly proportional to the degree of mitotic activity. For this reason the most rapid growth must be encouraged. A temperature of about 30° C is the best. In hard-coated seeds such as apples and pears, dissect out the embryo of seeds, freshly removed from the fruit before treatment. Under these conditions of most rapid growth the most severe treatment can be given. This will result in the highest proportion of cells in mitosis and the highest proportion in these of *regression* of anaphases, *i.e* of new tetraploid nuclei. Where treatment has been most effective growth is temporarily stopped and a swelling of the root-tips follows which is chiefly due to the upsetting of the polarisation of mitoses. The aim should then be to select the most retarded plants and nurse them back to normality by growing at high temperatures and under humid conditions. Rapidly growing annual plants are naturally easier to convert to polyploidy than slower growing perennials (Thomas 1945; for early references see Dermen 1940). A comprehensive review of the literature on colchicine has been made by Eigsti and Dustin (1955).

It should be noted that the action of drugs, unlike the effect of abnormal temperatures, continues long after treatment. For this reason prolonged treatment of seeds is often too effective and Thomas (unpub.) finds the maximum tetraploidy (80% of cells) in *Lolium* seedlings after 2 hours in 0·4% colchicine at 37° C.

Growth hormones have been used to stimulate mitosis in the normally non-dividing differentiated nuclei of roots and stems, in order to study the spontaneous occurrence of endopolyploidy and chromosome breakage in these tissues.

Table VI gives some examples of the reagents, treatments and organisms that have been used for these different purposes.

C. Differential Reactivity

Heterochromatic segments (H-segments) become visible with basic dyes as pale staining regions on metaphase chromosomes in somatic cells of certain plant species and Amphibia when they are chilled at 0–3° C for 3–5 days (Darlington and La Cour 1938, 1940, La Cour 1951, La Cour *et al.* 1956, Evans 1956, on *Trillium* etc., Callan 1942, on *Triton*). An explanation for this differential reactivity has been a matter of controversy for some years (Wilson and Boothroyd 1944, La Cour 1960, Boothroyd and Lima-de-Faria 1964, Haque 1963, Woodard and Swift 1964).

New observations in root tips of *Fritillaria lanceolata* suggest that in all examples of this phenomenon, as in this plant (La Cour and Wells 1973), the H-segments in thin sections (0·5 μm or less) stain less intensely than euchromatin throughout the whole mitotic cycle, irrespective or whether the plants are grown at low or moderate temperatures. This is because the comprising chromatin fibrils in this type of heterochromatin are at all times less compacted than those in euchromatin. The staining disparity is evidently masked according to stage, in squash preparations of root tips from plants grown at 18–20° C, by the thickness of the chromosomes and density of the chromocentres, respectively. The masking effect is alleviated at metaphase, however, when the plants are chilled at 0–3° C for 3–5 days, because the disparity in staining of the two kinds of chromatin is then enhanced by differential super-contraction of the euchromatin with chilling.

The differential reactivity to low temperature does not occur in *Scilla sibirica*, a species where the heterochromatin has a less open structure and remains condensed at telophase and throughout interphase (La Cour and Wells 1973).

Studies in mammalian cells have shown that heterochromatic segments may appear as extended regions in metaphase chromosomes, after incorporation of BUdR or 5-bromodeoxycytidine (BCdR) into DNA during the latter part of the S-period in the preceding interphase (Zakharov and Egolina 1972; Zakharov, Baranovskaya, Abraimov, Benjusck, Deminsteva and Oblapenko 1974).

The effect is further remarkable in that the differential reactivity is extended to a difference between sister chromatids, when the cells are kept in the presence of these thymidine analogues for two rounds of replication (Zakharov *et al.* 1974). This evidently occurs in cells where the DNA is then substituted in both strands of the helix in only one of the two sister chromatids.

It is not yet clear whether the effect created by such DNA substitution is a reflection of differential super-contraction in euchromatin, like that produced in some plants by low temperature (La Cour and Wells 1974), or actual de-spiralisation of the heterochromatic regions.

An uncertain degree of differentiation of heterochromatic segments, especially near the centromere, arises from X-raying and X-ray experiments can be combined with cold treatment in showing the relative effects of irradiation on euchromatin and heterochromatin (Darlington and La Cour 1945, Kurabayashi 1953).

This phenomenon has also been utilised for the recognition of *Trillium* chromosomes in *Paris quadrifolia* endosperm, present as a result of inter-generic fertilisation (Rutishauser and La Cour 1956). It was similarly ex-ploited by Japanese workers in studies of chromosome variation in natural populations of *Trillium kamtschaticum* (Haga and Kurabayashi 1954, Kurabayashi 1957) and of hybridisation and speciation (Haga 1956). In endosperm the weakly staining heterochromatic segments provide markers for studying the cytological consequences of crossing-over (Rutishauser 1955 *a* and *b*).

D. Temperature

Spindle formation is stopped below a critical temperature characteristic of each plant or animal: $3°$ C in *Triton*, $0°$ C in *Trillium*, $-5°$ C in *Fritillaria meleagris* (*cf.* Barber and Callan 1943). Low temperature can therefore be used as a substitute for drugs for all purposes mentioned except in birds and mammals). Callan (1942) with *Triton* has shown how temperature manipulation can be used for the joint control of spindle development, spiralisation and nucleic acid metabolism.

The same effect can be obtained by high temperatures (5 hours to 2 days at $30°–40°$ C) at mitosis and meiosis, *e.g.* PMC of *Fritillaria* (Barber 1940) and of *Trillium* (Matsuura 1937). Barber's work shows that heat, like X-rays and growth hormones, can throw the prophase nucleus back into the resting stage and thus give rise to diplo-chromosomes at the next mitosis. Temperatures in this range improve the clarity of prophase at meiosis in *Tradescantia* species (Swanson 1943; Dowrick 1957) and influence the number and position of chiasmata in *Tradescantia bracteata, Uvularia perfoliata* (Dowrick 1957) and *Hyacinthus orientalis* (Elliott 1955). In *Lolium perenne* treatment induces differential condensation in bivalents at meiosis (Jain 1957).

Higher temperatures (*ca.* $44°$ C) for a short period have the same effect by shock and have been used on young embryos for producing polyploid cereals (Randolph 1932, Dorsey 1936) and *Linum* (Lutkov 1938). Before meiosis they reduce chromosome pairing (Straub 1936). Their detailed action on chromosomes and spindles has been analysed by Barber and Callan (1943).

In self-incompatible plants unreduced (diploid) pollen grains, produced by heat-shocks which inhibit one of the meiotic divisions, will accomplish self-fertilisation. By applying this method to self-incompatible apples and pears Lewis (1943) has accordingly been able to produce seedlings with uniformly triploid chromosome numbers.

E. Centrifuging

Experiments have been made for various purposes by Beams and King (1935), Kostoff (1938 *b*) and Kawaguchi (1938).

F. Various Genetical Devices

Most combinations of chromosomes and breeding work are too obvious or too elaborate to be given here. Five intermediate hints may be mentioned.

(1) With plants it is usually easy to take out one anther and later examine or breed from another. It is possible to take roots of seedlings so as to decide from which to breed.

(2) In animals, special procedure is often necessary, *e.g.* in mammals to remove one testis for examination and leave the other for breeding.

(3) In *Drosophila* the progeny of an important parent are examined (in the salivary glands) while the parent is still alive, in order to discover its chromosome make-up before using it for experimental breeding. Where the presence of the Y chromosome is important, mitosis in the oesophageal ganglia must be used (acetic-orcein smear).

Ganglia should be kept in 70% alcohol for parallel test with salivary gland in case they are required in breeding experiments.

(4) The genetic activity of chromosomes can be studied most rapidly and most conveniently in pollen, for very anther of a hybrid plant contains a genetic experiment. A simple study of this kind is described by Levan (1939*b*), a more complicated one by Darlington and Thomas (1941). The genetic implications of embryo-sac studies are reviewed by Darlington and La Cour (1942).

(5) There are two basic genetical methods of inducing change in genes and chromosomes of all plants and animals. They are inbreeding and out-breeding. It is by inbreeding that a variety of specific abnormalities of chromosome behaviour have been induced in rye, maize and other plants (*cf.* Beadle 1933, Rees 1955). Outbreeding, on the other hand, that is the crossing of species, is most notable for its effects in *Triturus* (Lantz and Callan 1954) and *Allium* (Emsweller *et al.* 1935).

10

The Control of Fertilisation

A. Pollen Germination

Artificial germination of pollen is necessary for two purposes: the measurement of fertility and the observation of the generative nucleus in the pollen tube. Free growth of pollen tubes demands special culture media with temperatures controlled at about 20° C and humidity approaching saturation. Calcium ions have been found to be an important factor in promoting optimum growth, but for maximum utilisation apparently require a medium of low hydrogen-ion concentration which contains magnesium and potassium salts (Brewbaker and Kwack 1963, Kwack and Kim 1967).

Culture Media:
(1) Cane sugar solutions in water of concentrations 3% to 30% plus 0·03% calcium nitrate, 0·02% magnesium sulphate and 0·01% potassium nitrate; adjust to pH 8·3 (Kwack and Kim 1967).
(2) The same (*a*) in agar (*b*) in 2% agar, 2% gelatin (Newcomer 1938).
(3) Cane sugar solutions with extract of style and placenta, or stigmatic secretion (Yasuda 1934).
(4) Stigmatic secretion alone (Lewis unpub. *Prunus, Pyrus*).

The percentages of sugar solutions giving optimum speeds and chances of germination vary with the species, the individual, and no doubt even the pollen grain itself. They also vary with the temperature of germination (rising with a fall of temperature) and with the conditions, particularly of temperature at which the plants are grown. Sudden changes of temperature cause pollen grains in wet or humid conditions to burst. The following list will in these circumstances provide a rough guide to sugar concentration at about 20° C:

Papaver sp., 3%, Beatty 1937
Eschscholtzia sp., 5%, Beatty 1937
Tradescantia 2*x*, 8%, Swanson 1942
Scilla spp., 15%, Gagnieu unpub.
Trillium grandiflorum, 15%, Lewis unpub.
Primula obconica, 10–20%, Lewis unpub.

Amaryllis belladonna, 8%, Newcomer 1938
Tulipa spp., 14%, Upcott 1936*b*
Prunus avium, 20%, Lewis unpub. (10% at 30° C)
Pyrus spp., 20–30%, Lewis and Thomas unpub.
Solanum Lycopersicum, 2*x*, 25%, Lewis unpub.

Observation with controlled humidity is simplest by the hanging drop method. A drop of the medium containing the pollen is placed on a cover slip which is inverted over a ring (Fig. 3*g* App. iv) smeared with petroleum jelly to make a hanging drop chamber. The humidity can be increased by adding a drop of water on the slide, or by placing a drop of agar beside the hanging drop on the cover slip. A more accurate method is to leave the ring unsealed and place the slide in a desiccator with glycerine in water, adjusted by trial to give the correct humidity: 20% glycerine gives 50% humidity at 15° C.

Rapid method for germination test. Dust the dry pollen on a cover slip, invert over hanging drop chamber containing a small piece of wet filter paper. Adjust the size of the paper to give correct humidity (Thomas unpub.).

B. Tube Division

In Angiosperms (apart from Gramineae) the generative nucleus usually divides into the two sperm nuclei in the pollen tube. In *Tradescantia* and *Tulipa* the division is 12–24 hours after germination at room temperature. A cover slip with the pollen germinated as already described can then be stained and fixed either by a stain-fixative or by 2BE–CV (Upcott 1936).

Recent improvements make this mitosis one of the most favourable for the study of chromosome shape, structure and breakage. The coated slide technique (Conger 1953), floating cellophane method (Schedule 23) and floating membrane technique (Savage 1957) offer satisfactory alternative methods. In all three a mitosis-arresting agent can be used to accumulate metaphases provided that mitosis does not take place too quickly. In *Tradescantia* the interval is satisfactory, being about 16 hours at 20° C (Swanson 1940). This method can be further combined with the study of X-ray effects (Darlington and La Cour 1945, Bishop 1950) or that of oxygen (Conger and Fairchild 1952).

C. Pollen Storage

For purposes of cross-fertilisation it is often necessary to store pollen. Ripe anthers are allowed to burst in petri dishes and the pollen is stored in vials stoppered with cotton wool, which are placed in desiccators with humidity controlled at 50% by using glycerine in water. A list of the storage capacities of 500 species, and of sugar concentrations they require for germination, is given by Doroshenko (1928; see also Visser 1955). Kept at 0–8° C pollen has been found to maintain its capacity for germination for long periods. Where longer storage is required, freshly collected pollen from some plants may be stored over anhydrous calcium chloride at 4° C for one year (Kihara 1919), or for periods more than twice as long in vacuum storage (Jensen 1964). Evidently storage in organic solvents such as anhydrous acetone also provides a means of maintaining pollen viability (Iwanami and Nakamura 1972). The freshly collected pollen is initially kept dry over silica gel at $-10°$ C.

Although with long storage by the above methods pollen may maintain its capacity for germination, it may not maintain its capacity for fertilisation so long (Nebel and Ruttle 1939).

In this regard pollen seems to be of two kinds. In grasses the generative nucleus divides into two sperms before germination. This pollen loses its capacity for germination in about 40 days, for fertilisation in about 1 day. In other plants whose generative nucleus divides after germination in the pollen tube, the pollen will germinate after two or three years, and it loses its capacity to fertilise only in 40–200 days. The failure of fertilising power is probably due in grasses to the death of the sperm nuclei; elsewhere to the fusion of sister chromatids shown to occur in the generative nucleus after storage (Barber 1938).

Dead pollen such as that of old herbarium specimens can still be examined by acetic stains in order to discover its size and quality and the numbers and positions of the vegetative and generative nuclei (Darlington and Thomas 1941; Celarier and Mehra 1958).

D. The Style

Fertilisation in plants is controlled in the first place by the style-pollen relationship, which has been the subject of extensive genetic and cytological study in regard to both self- and cross-sterility. Styles examined from four hours to six days after pollination (according to the rate of growth) show differences in the progress of pollen tubes according either to their absolute and relative genetic constitutions, or to the temperature, or to a special effect of irradiation. Thus X-rayed styles allow incompatible pollen to grow in special crosses of *Triticum* according to Tanaka (1937) and in self-pollination of *Oenothera* according to Lewis (unpub.).

Treatment depends on the size of the style. Large styles usually have thick central strands of conducting tissue. These strands contain all the pollen tubes and they can be removed intact with the stigma after the style is slit (*Datura*, Buchholz 1931; *Oenothera*, Emerson 1938).

In small styles the central strand is more fragile and special methods must be used to get at it, according to its size and shape:

(1) *Material fresh or fixed* 10 *min in acetic alcohol* and stored if necessary in 30% alcohol to prevent hardening.

(i) Cut the style in half (*Pyrus, Prunus, Primula,* Lewis and Modlibowska

1942). Crush afterwards under the cover slip (*Tradescantia*, Anderson and Sax 1934; *Secale*, Sears 1937).

(ii) Boil in 4% sodium sulphite solution for about 3–10 min and press out the central strand under a cover slip (*Pelargonium, Petunia, Nicotiana*, etc., Sears 1937).

(iii) Section on freezing microtome at 20 μm (plants with straight styles, *e.g. Prunus* and *Primula*, Lewis unpub.).

Stains for (1):

(a) Acid-Fuchsin with Light-Green, 5 min to 6 hr at 55° C (*Primula, Pyrus* and generally).

(b) Cotton Blue, 5 min to 6 hr at 55° C (Cereals, Watkins 1925, *Prunus, Oenothera*).

(c) Lacmoid—Martius Yellow (*Pyrus*, Nebel 1931).

(d) Fluorochrome from Water Blue, giving fluorescence in ultraviolet light (various species, Linskens and Esser 1957, Martin 1959, Kho and Baer 1968).

(2) *Material fixed in Medium Flemming*

(e) Embed and stain in Delafield's haematoxylin or (a) (b) (c).

E. Haploid Plants

Haploid plants are important for the light they throw on the genetic constitution of their parents; for the study of their chromosome pairing at meiosis; for the production of homozygous diploids by their later doubling; and for making species crosses possible.

Haploid parthenogenesis can probably be induced in most diploid and even-numbered polyploid plants (4*x*, 6*x*, etc.), inbred diploids being easier than outbred. Certain special conditions favour this development (*cf*. Darlington 1937):

(1) Pollination with another species (*Datura, Solanum, Nicotiana*).

(2) Pollination with a plant of a different ploidy (*Zea, Campanula, Petunia*).

(3) Delayed pollination giving rise to twins (*Triticum*, Kihara 1940, L. Smith 1946).

(4) Use of X-rayed pollen (*Triticum, Nicotiana*, Kihara and Yamashita 1938, L. Smith 1946).

(5) From pollen by anther culture (Bougin and Nitsch 1967, Nitsch and Nitsch 1969, Sunderland and Wicks 1971).

Among natural conditions, twinning has been found to lead to the development of haploids or triploids, or both, in *Triticum, Poa, Gossypium, Linum* and *Solanum* (*cf*. Harland 1936; Skovsted 1939).

F. Haploid Animals

The development of haploid eggs following partial or complete suppression of doubling of fertilisation has been studied chiefly in Echinoderms and Amphibia (*cf*. Wilson 1937). The object is to trace the relationships of chromosomes, mitotic spindles and general development.

Eggs can be made to undergo incomplete cleavage without fertilisation or with multiple fertilisation by (i) shaking or pricking, (ii) placing in hypertonic solution, (iii) treatment with drugs such as strychnine.

Development of egg fragments without egg nuclei can be induced in *Triton* by separating a part of an egg containing the sperm by ligature (Fankhauser 1937, Fankhauser *et al*. 1952). Comparable experiments have been successful in studying fusion in *Paramecium* (Tartar and Chen 1941).

In the Hymenoptera, where the control of fertilisation is a necessary part of the breeding system, many factors, internal and external, control it (*e.g.* temperature, R. L. Anderson 1936).

11

Photography

A. Uses

Microphotos can show either less or more than drawings according to how they are chosen. If ill chosen they will certainly show less, and if they require arrows and signs to explain their meaning they must be regarded as ill chosen. The first problem in microphotography is therefore to choose a subject capable of illustration in this way. Where the subject is unsuitable it can be made suitable in five ways:

(1) By using flat smears and squashes instead of sections.
(2) By adapting the method of fixation so as to give the flat subject required.
(3) By using a deeper or more selective stain.
(4) By using an appropriate colour filter either to enhance or reduce contrast, (see Table VII) or to more readily separate the stained image of chromosomes, nucleoli etc. from overlying silver grains in autoradiography.
(5) By using a lower power oil-immersion objective to give greater depth of focus.

The best general references to microphotography are: Shillaber 1944, Stevens 1957, Michel 1957, and Needham 1958.

B. Cameras

In recent years cameras using plates have, somewhat regrettably, been largely superseded by cameras employing 35 mm film.

A 35 mm reflex camera without the lens can be adapted to take microphotos. The same camera with 'microneg' film and with its lens can be used for copying manuscripts. Camera bodies, minus a lens, for use with 35 mm film and incorporating a shutter as well as a focusing device, operative between the eyepiece of the microscope and focal plane, can be obtained commercially. In some more elaborate commercial systems a cassette for 35 mm film is inserted in the body of the microscope, correct focusing and exposure being virtually automatic.

C. Taking the Photograph

(1) *Preliminary*

(i) The illumination and the cleanliness of the apparatus naturally require more attention for photography than for casual use. The extraneous window light must be cut off.

(ii) The colour filters will depend partly on the type of emulsion of the negative material used, as well as the stain and degree of contrast desired in the print. Plates or films with fine grain panchromatic emulsions and a choice of medium or high contrast are best employed. A filter will transmit its own colour and absorb its complementary colour (Table VII), so that in practice negatives can be produced giving lighter or darker images on the print, respectively.

A neutral filter can be used to reduce the intensity of light.

(iii) For all purposes a partial cone of light is the best. The substage diaphragm should be cut down, but not to the point of producing refraction images or optifacts (Darlington and La Cour 1938).

(iv) The bench must be set up to avoid vibration.

(v) The slide can be tilted to flatten the object where a slight correction is needed (White unpub.). Pollen grains in liquid preparations may be rolled by sliding the cover slip.

(vi) The eyepiece should be of suitable power to accommodate the image within the 35 mm frame. A special projection eyepiece is unnecessary.

(2) *Operative*

The correct length of exposure must be determined by trial and error; it may vary by as much as 10 seconds to 2 min according to the illumination, filters and speed of the emulsion. It varies as the square of the linear magnification. Exposure meters are now available for microphotography.

D. Developing

It is convenient to be able to enlarge photographs up to five times for publication and up to ten times for demonstration pictures. Usually for publication the magnification should seldom exceed ×2000. Prints of this scale of enlargement, with a minimum of grain, can readily be obtained from negatives with present-day emulsions recommended for microphotography, provided they are developed correctly with the manufacturer's suggested developer or suitable substitute. Correct development requires that the films are developed for the recommended time at a constant temperature, usually 20° C.

The stop bath (2·5% acetic acid), fixer and, ideally, the water for washing should be used at the same temperature as the developer. After washing, films should receive a final rinse in distilled water, in order to avoid water marks. The processed plates and films are best dried in a drying cabinet, away from dust.

Only as a makeshift should after-treatment of the negative be used as a remedy for imperfect exposure. Potassium ferricyanide reducer (see App. II) can be used for correction of over exposure, but tends to increase contrast. A uranium type intensifier is possibly best for slight under exposure.

E. Printing and Regulation of Contrast

The test of a good print is that it should show as much as the negative. The better the negative the easier this is.

Bromide paper is made in a series graduated in capacity between hard contrast and soft contrast. The best negatives usually require a medium grade of paper, but your choice must depend to some extent on what you want to show. Avoid over-contrasted prints, because with copying for publication contrast will be increased yet again.

TABLE VII – PHYSICAL PROPERTIES OF FILTERS

Filter	*Absorbs*	*Transmits*
Red	Blue and Green	Red
Green	Red and Blue	Green
Blue	Red and Green	Blue
Yellow	Blue	Red and Green
Cyan	Red	Blue and Green
Magenta	Green	Red and Blue

The lighting system in the enlarger has some effect on contrast. A point source gives high contrast: a diffuse source slighter contrast. Compensation for slight inequalities in the negative *e.g.* uneven density resulting from uneven illumination, or over-density in the nucleolus etc., can sometimes be corrected by suitable shading during exposure of the print at the time of enlargement.

F. Preparing Photographs for Reproduction

Two methods of reproduction are in common use by scientific publishers: half-tone (as in this book) and collotype. Both these methods are more expensive than the zinc block used for line reproduction of pen-and-ink drawings.

For reproduction, glossy paper is necessary, and the prints are glazed by rolling and drying on the chromium-plated sheet of a commercial glazer, or if unavailable on rigorously cleaned plate glass.

Since the photograph should, as far as possible, explain itself, irrelevant structures should be excluded by vignetting in printing.

Some publishers prefer photographs unmounted. If mounted, the photograph or group should leave no gaps or patches but should fill the whole plate as nearly as possible. In one plate all the photographs benefit by having a uniform intensity, background, and magnification.

G. Screen Projection

For demonstration of chromosomes, in teaching, it is a useful asset if images of cells can be projected either by simple screen projection (Darlington and Osterstock 1936), or by closed-circuit television. Either method allows the focus and position on the slide to be varied while both demonstrator and audience have the image in view. A more powerful light source than is ordinarily used for microscopy is required. A heat absorbing filter may also be desirable.

12

Autoradiography

A. Uses

Autoradiography is a photographic method for locating and measuring radioactive inclusions in a specimen. Radioactive isotopes (C^{14}, P^{32}, S^{35}, etc.) are introduced into the plant or animal by way of specially constructed molecules. This method has been applied to the study of organs, tissue cells and cell organelles. It permits the indirect investigation of metabolic processes within cells and tissues thus extending the scope of cell chemistry.

With the introduction of tritium, autoradiography has provided significant information on the replication of chromosomes, time and place of synthesis of the two nucleic acids in mitotic and meiotic chromosomes, the metabolic activity of the nucleolus as well as that in puffs of giant polytene chromosomes. Some selected references to such studies are given in Table VIII.

The physical principles of autoradiography and their application to cell chemistry have been discussed by Pelc (1958), and the quantitative aspects by Pelc (1958) and Taylor (1956).

B. Methods

The material is prepared in smears or squashes, or sections cut at 4–6 μm. It is then coated with a photographic emulsion sensitive to the ionising radiation emitted by the active atoms and left in the dark at 4° C for a few days before photographic processing.

Three methods of preparation are suitable for cytochemistry:

(1) Bêlanger and Le Blond's (1946) method. Melted photographic emulsion is used to coat the section. The method was modified by Bêlanger (1950) to allow the section to be stained after processing. Emulsions of varying grain size can be applied in different thicknesses which allow some flexibility. A variation of the technique for use with micro-organisms has been described by King et al. (1951) and Levi (1954).

(2) Evan's (1947) and Endicott and Jagoda's (1947) method. Paraffin sections of tissues are floated on a photographic plate or film. After being dried and left for exposure, the paraffin wax is removed before processing. The resolving power of this technique is relatively poor, owing to the thickness of the usual photographic materials.

(3) The stripping-film technique of contrast autoradiography (Pelc 1947, Doniach and Pelc 1950, Pelc 1956; Schedule 22). In this widely used method, which is suitable for chromosome studies, a piece of stripped film, consisting of a bottom layer of pure gelatine and a thinner upper layer of nuclear track emulsion, is floated emulsion side down on to the slide carrying sections or smeared cells, after expansion of the film in distilled water or sucrose–potassium bromide solution at 18°–21° C. The filmed slides are then dried and stored in darkness at about 4° C until processing is desired.

An appraisal of autoradiographic techniques and recent developments is given by Rogers (1967).

C. Preparation of Tissues

Sections. These are prepared in the usual way, but the choice of fixative is limited since artefacts such as fogging or desensitisation of the emulsion can be produced. Alcohol, acetic-alcohol, formol-acetic-alcohol, formol saline, freeze-substitution methods (*cf.* Taylor 1956) are satisfactory. Some other fixatives may perhaps be used if thorough washing is observed or if the sections are coated with celloidin before setting up the autoradiographs. The preparations should however be checked for artefacts.

To ensure adherence of film to glass it is necessary that the slides used for mounting sections should be 'subbed', that is filmed with egg albumen or dipped in a 0·5% aqueous solution of gelatine and 0·1% chrome alum. Dry the subbed slides for at least two days before using, but on no account attempt to hasten drying by heating over a flame. It is worth while drying slides in an incubator at 37° C or less since this avoids contamination by dust. Preferably, slides cleaned only in alcohol should be used in autoradiography. Some makes of glass produce a high background, if the background over the glass is high different makes should be tested.

Smears. These are prepared in the usual way, but note on subbed slides. The choice of fixatives is again limited to those recommended.

Squashes. Roots, anthers, testes, etc., are best fixed in acetic-alcohol. The tissues can be hydrolysed for 6–8 min in N HCl at 60° C to facilitate squashing, or plant tissues can be treated with pectinase (Ch. 5) to avoid loss of RNA or adenine from RNA. Small pieces of tissue are tapped out in 10% acetic acid, in order to separate cells as a single layer, preferably on a cover slip which is then inverted on to the subbed slide. Flatten the cells by slight pressure on the cover slip but not with finger tips in case they become contaminated with radioactive material. Salivary glands should be squashed on subbed slides in 45% acetic acid. The prepared slides must not be heated. After squashing they are best left on the bench for 5–10 min before inverting the slide in a ridged dish containing 10% acetic acid to separate the slide from cover slip.

Staining. Sections and smears may be stained in Feulgen before setting up the autoradiographs, similarly Feulgen squashes can be utilised for auto-radiography. If tissues however are stained in this way the prepared slides must be thoroughly rinsed several times in distilled water before setting up the autoradiographs.

A limited number of stains are available for staining preparations after photographic processing (see Schedule 22). With most stains the gelatine takes up some stain which results in a loss of brilliance. Toluidine blue is very suitable for staining nuclei if the tissues have previously been hydrolysed in N HCl. 'Euparal' is suitable for mounting stained preparations.

D. Exposure of Autoradiographs

During storage developable photographic grains are produced in the super-imposed emulsion by ionising radiations emitted from the sites of radio-activity in the tissues. The exposure time required depends on the isotope used and its concentration in the tissue. It varies from a few days to several weeks.

E. The Amount of Tracer Necessary to Obtain Autoradiographs

A certain minimal concentration of labelled compound or tracer is necessary to produce an observable autoradiograph. Too heavy an application has to be avoided, since metabolic processes may be affected either by the chemical applied or by the radiation emitted by the tracer. A method of calcula-ting the minimal concentration required is described by Pelc (1951; *cf.* 1958). In determining the amount of labelled compound to be used the following factors have also to be considered: (i) purpose of investigation; (ii) radio-activity decay between application of tracer and application of film to tissues; (iii) the fate of the applied chemical compound in the organism and its possible extraction during preparation.

In autoradiography it is obviously important to use precursors which are as specific as possible for the chemical component being investigated. It is equally important to obtain autoradiographs from a series of fixations at different times after application, since it cannot be assumed that incorporation is always rapid or that compounds are necessarily synthesised where they are located by conventional cytochemical methods. In respect of timing, the control of temperature is especially essential in plants. Some indication of the amount of labelled compound to use and required exposure times can be gained from a survey of previous cytochemical applications of autoradiography (*cf*. Pelc 1958).

F. Background and Artefacts

Before setting up autoradiographs it is necessary to remove any tracer which may remain in the acid-soluble precursor pool within the tissues. Specimens on the slides should therefore be steeped in ice-cold 5% trichloroacetic acid for 15–30 min, according to the acidity of the fixative in which the tissues were fixed. Obviously such treatment is unnecessary if the tissues are to be hydrolysed for staining in Feulgen. Whatever treatment is employed prior to autoradiography the preparations must be thoroughly rinsed in distilled water to remove contaminants which may give background.

Background. It is impossible to obtain autoradiographs which are completely free from background since a number of grains develop without exposure to radioactive materials. Background increases with age of the film and is increased by over-exposure to the safelight. Direct illumination from the safelight is best avoided when setting up the autoradiographs and especially during their development. Background may also be increased by static electricity during the stripping of film from the plate. Stripping plates should be fresh and are best kept at 4° C, and obviously, away from sources of penetrating radiation and chemical fumes. If the exposure time is expected to be longer than two weeks, a transfer solution of sucrose and potassium bromide should be used instead of water.

The degree of background in so far as it affects interpretation is discussed by Pelc (1958) and Taylor (1956).

Artefacts. Artefacts are generally due to the diffusion of substances from the specimen which act on the photographic emulsion making grains developable. They can generally be distinguished from the grains produced by ionising radiation since they appear as a single layer next to the tissue. It is wise, however, to have controls without radioactive tracer since these will always reveal the same artefacts as the active specimen. This is particularly necessary when using new experimental organisms or tissues, new staining techniques before preparation of autoradiographs and fixatives other than those recommended. Care in handling autoradiographs, cleanliness in the darkroom and in the handling of slides, are all factors in success.

G. Observations of Autoradiographs

The preparation of autoradiographs is described in Schedule 22). Observation can be made either of unstained or stained preparations, the choice depending on the circumstances. Relatively small differences in the autoradiographs are more readily noticed in unstained preparations, as are some artefacts such as pigments or granules in tissues. Stained preparations are preferable if accurate localisation is desired.

Unstained preparations can be permanently mounted in a special medium (see Formulae of reagents) or mounted temporarily for inspection by flooding with distilled water. With care, temporary mounting can be repeated many times without damage to the autoradiograph, and preparations can be stained later if desired. Unstained preparations are best observed under a phase-contrast microscope. The tissues can first be viewed by phase and the superimposed autoradiograph next by removal of the annulus in front of the condenser.

H. Removing an Autoradiograph

In certain circumstances it may be desirable to strip an autoradiograph from the slide *e.g.* (i) To apply fresh stripping film in order to obtain another autoradiograph exposed for a different length of time.

(ii) To obtain a microphotograph of labelled chromosomes unobscured by silver grains.

(iii) To re-stain the chromosomes with a different stain.

A simple but often effective method for sections is to steep, after removal of mountant and cover slip, the slide for 5–10 min in distilled water at 30° C, the film can then be gently lifted off with fine forceps without disturbing the sections. Alternatively, the dry processed film can be removed from the slides in 45% acetic acid (Das and Alfert 1963). A more elaborate method involves potassium ferrocyanide to dissolve out the silver grains and trypsin to digest the gelatin of the film (Bianchi *et al.* 1964). It may be noted that intense silver staining of nucleoli sometimes occurs in preparations as a result of making repeat autoradiographs (Das and Alfert 1963).

TABLE VIII – SCOPE OF AUTORADIOGRAPHY

Organism	Tissue	Label	Purpose	References
Lilium longiflorum	anthers	P³²	time of DNA synthesis in relation to meiosis	Taylor and McMaster 1954
Vicia faba	root tips	P³²	time and duration of DNA synthesis	Pelc and Howard 1952
Vicia faba	root tips	H³-thymidine	chromosome replication	Taylor et al. 1957
Vicia faba	root tips	H³-thymidine	chromosome replication	Peacock 1963
Vicia faba	root tips	H³-thymidine	patterns of DNA replication	Evans 1964
Bellevalia romano	root tips	H³-thymidine	sister chromatid exchange	Taylor 1958
Secale cereale	leaves	H³-thymidine	DNA synthesis: euchromatin versus heterochromatin	Lima-de-Faria 1959
Melanoplus differentialis	testes	H³-thymidine	DNA synthesis: euchromatin versus heterochromatin	Lima-de-Faria 1959
Drosophila melanogaster	salivary glands	H³-thymidine	ordered replication of DNA	Plaut et al. 1966
Triturus vulgaris	testes	H³-thymidine	time course of meiosis	Callan and Taylor 1968
Chironomus	salivary glands	H³-uridine	RNA activity in puffs	Pelling 1959, 1964
Schistocerca gregaria, Cyratacanthacris tartarica, Chorthippus brunneus	testes	H³-uridine	RNA synthesis at meiosis and spermiogenesis	Henderson 1964

Triturus vulgaris	oocytes	H³-uridine	RNA synthesis in lamprush chromosomes	Gall and Callan 1962
Vicia faba	root tips	H³-adenosine	RNA in metaphase chromosomes	La Cour 1963
Trillium cernuum				
Zea mays	anthers	H³-uridine and H³-cytidine	synthesis of nucleolar RNA at meiosis	Das 1965
Drosophila virilis & *D. melanogaster*	salivary glands	H³-leucine and S³⁵-methionine	labelling of chromosomal protein	Sirlin and Knight 1960
Xenopus laevis	oocytes	H³-r RNA	detection of sites of r RNA–DNA hybridisation	Gall and Pardue 1969
Plethodon c. cinereus	testes	H³-RNA	detection of RNA–DNA hybrid molecules in mitotic and meiotic chromosomes	MacGregor and Kezer 1971

13

Describing the Results

A. Interpretation

When we have our preparations fixed and stained, how are we to know the value of what we see in them?

Clearly the first problem is the construction of a three-dimensional model in the mind's eye from a continuous series of focal projections seen by the physical eye. This is a question of training, and probably trainability. It must be left to practice, especially practice in drawing, to solve.

The second problem is more intellectual, that of validity. Obviously if we can compare fixed with living observations we have a direct test of validity. This test has now been greatly extended by the use of ultraviolet photography. But this method is restricted by the elaboration and costliness of the apparatus. It has therefore been applied so far only to special studies, which have shown the general soundness of the images obtained by what we earlier recognised as 'good fixation'. This recognition was made by comparisons of other kinds, namely:

(1) Of chromosome behaviour in similar cells of different species.

(2) Of chromosome behaviour in the same cells with different treatments and

(3) Of the succession of observations during development which provide a satisfactory basis of prediction.

Interpretation therefore depends on common sense and on taking into account all the relevant data. These data may now extend beyond the description of the chromosomes. They include the results of the new methods of experiment already described. They include also considerations of genetical experiments and protein chemistry, reinforced by X-ray analysis (Bernal 1940). This extension of the critical apparatus of cytology has gradually removed the main grounds of contention and has helped to establish a wide measure of agreement in the interpretation of chromosome behaviour.

B. Illustration

Drawing is an integral part of chromosome research. It is doubtful whether good training or even good observation is compatible with bad drawing. But it is certain that its value is largely wasted if unsubstantiated by good drawing. What is needed is a drawing which (i) can be reproduced cheaply, (ii) will show the reader, true or false, exactly what the draughtsman thought he saw. For these purposes pen-and-ink drawing alone is suitable.

As the needs of research change, the subjects and manner of illustration must change too, and a great deal will depend on their adaptability. For this everything hinges on the necessary detail of the subject in question. The original magnification of drawings should always be about half again as big as the smallest size at which they will show this necessary detail, and at which they should be reproduced. To allow for this reduction all lines and all spaces between lines should usually be bold, *i.e.* about 0·5 mm thick.

Again, in different drawings of the same subject, the purpose and there-fore the significant elements will change. Sometimes it is the arrangement of the chromosomes in the cell and sometimes the structure of individual chromosomes. In the first case the unit of drawing is the whole cell, in the second each chromosome can, and perhaps should, be drawn separately. Sections of course, where they can be complete, have the advantage, for the first, smears and squashes for the second. Particular chromosomes can be picked out by showing them solid or cross-hatched or stippled as against others in outline. Further, everything should be done in the way of lettering and marking special structures to enable the reader to see at once what is referred to in the texts. Finally, where the case is novel and intricate, diagrams and graphs should be used to clarify the description or explain the hypothesis.

And as to good training, it means simply this: take care of the drawing at the beginning and in the end the drawing will take care of itself. A slovenly start cannot be made good. Much can be learnt from the principles of line illustration described by Staniland (1952).

C. Description

When the drawings and photographs are prepared, and the data of observa-tion collected, the work is ready for description. Discoveries unpublished are discoveries wasted, and the author himself will increase the accuracy of his observation by submitting to the discipline of description. This is a necessary part of technique, and one which has made some advances recently, habit and tradition notwithstanding. Certain rules are therefore worth setting out:

First, assess your results in relation to the previous work of others. Consider everything that is relevant, whether or not it arises by similar methods or from similar organisms. For relevance arises in many ways, and how to discover them is all the more important in research for its being neglected. Never fear to make surprising conjunctions of plant and animal, of genetics and physiology, for it is of the nature of chromosome studies to bring such conjunctions about. Do not, on the other hand, quote papers which you believe to be worthless, unless you wish to prove them so. And in proving them so, never say that the proof is absolute or certain for that, as everyone knows, only time can show.

Secondly, be concise. This you can do in many ways:

§ Make your title short. Do not offer a 'preliminary account', for all scientific work is preliminary. Do not put Linnaeus or Pallas in the title; leave them to enrich the text. State the specific and analytical result rather than the general and superficial class of your observations.

§ Convey information in your summary. Let it be exact without detail and general without vagueness. Do not say 'A, B and C were discussed'. Give your conclusion if you reached any. There are some who will tell you to arrive at no conclusion, for it will only 'weaken your argument'. Do not believe it. If you are unable to state any conclusion, leave out the discussion, for not even the professional abstractors will read it.

§ Obtain and give quantitative results where a sharp classification makes it possible: provided of course that some useful conclusion can be drawn from them. Never say merely that such an event was 'frequent' or 'very frequent', since that means nothing to anyone else. If you do not know precisely, give an estimate.

§ Put all possible facts in tabular form since if you do not, your readers will have to do so, and their number will be fewer. On the other hand do not offer tables which mean nothing.

§ Use technical terms consistently and state your definitions of terms that have not been defined. Do not 'furnish data' – just state the facts. Do not 'describe phenomena' – just say what you saw. Avoid vocables like karyokinesis, arrhenotoky, megagametophyte, photomicrography. They exist only in books, and in the memories of over-learned men.

§ Omit personal matters. Let your results alone speak for your exertions. Especially take care not to explain why you failed to get results. The reader is selfish. He is interested only in your success.

In a word, pay attention to the reader if you wish him to pay attention to you.

APPENDIX I

Sources of Material

In order that the unity of chromosome behaviour can be turned to advantage we have collected the following references to the most recent papers describing the handling of chromosomes in various classes and families (Table VIII) and in relation to the time of year (Table IX).

ABBREVIATIONS

B Chrs.	supernumerary chromosomes	G. bandg.	Giemsa banding
Dim.	diminution	Hap.	haploid
Emb.	embryo	Hyb.	hybrid
End.	endosperm	Mei.	meiosis
Exp.	experimental	Mit.	mitosis

Non-loc. C.	polycentric (or diffuse)		
Parth.	parthenogenesis		
Ppd.	polyploid		
Sp.	spindle		
Xta.	chiasmata		

TABLE IX – CHROMOSOMES OF NATURAL GROUPS

Group	Genus, etc.	Subject	Author
PROTISTA (general)	—	—	Grell '52
Flagellata	*Trichonympha*	Mit., Sp. Mei.	Cleveland '49
Flagellata	*Euglena*, etc.	Mit.	Leedale '58
Radiolaria	*Collozoum*	Mit., Sp.	Pätau '37
Radiolaria	*Aulacantha*	Endomitosis	Grell '53
Foraminifera	*Rotaliella*	Meiosis	Grell '54
Ciliata	*Paramecium*	Conjugation	Chen '46
Ciliata	*Zelleriella*	Mitosis	Chen '48
ANIMALS			
1. PLATYHELMIA			
Turbellaria	*Dendrocoelum*	Mei. ♀	Gelei '21
Turbellaria	*Mesostoma*	Xta. B Chrs.	Husted *et al.* '40
Turbellaria	*Dugezia*	Mei. Parth.	Benazzi '57
Trematoda	*Schistosomum*	Xta.	Ikeda *et al.* '36
2. NEMATODA	*Ascaris*	Ovum, Dim.	White '36, Lin '54
3. ANNELIDA			
Polychaeta	*Tomopterus*	Mei. ♂	Schreiners '06
Oligochaeta	*Eisenia*, etc.	Mei. ♀ Parth. Ppd.	Muldal '52
4. MOLLUSCA			
Gastropoda	*Murex*	Mei. ♀, etc.	Staiger '51
Gastropoda	*Triodopsis*	Mei. ♂	Husted *et al.* '46
5. ECHINOIDEA	*Urechis*	Mit. Parth.	Belar '33

6. ARTICULATA

Crustacea

(1) Phyllopoda	*Daphnia*	Mei. ♀, Parth.	Ojima '58
(2) Copepoda	*Ectocyclops*	Mei. ♀, Emb.	Beermann '54
(3) Amphipoda	*Gammarus*	Mei. ♀, Emb.	Brian and Callan '57
(4) Isopoda	*Anilocra*	Mei. ♂, Xta.	Callan '40
(5) Ostracoda	*Heterocypris*	Ppd. Parth. B Chrs.	Bauer '40

Arachnida

(1) Scorpiones	*Tityus*	Mit. Mei. Non-loc. C.	Brieger and Graner '43
(2) Acarina	*Pediculoides*	Mei. Parth. Hap.	Pätau '36
Acarina	*Pediculopsis*	Ovum, Xta.	Cooper '37
(3) Araneae	*Aranea*	Meiosis	Pätau '48

Myriapoda

	Otocryptops	Sex Chrs.	Ogawa '54

Insecta

(1) Neuroptera	various	Mei. ♂	H. O. T. Suomalainen '52
(2) Odonata	various	Mei. ♀, Xta.	Oksala '54
(3) Orthoptera	Mantidae	Mei. ♂	Callan and Jacobs '57
Orthoptera	Mantidae	Mei. ♂, Sex Chrs.	White '65
Orthoptera	*Anacridium*	Mei. ♂, ♀	Colombo '54
Orthoptera	*Periplaneta*	Hyb. Mei. ♂	John and Lewis '57
Orthoptera	*Schistocerca*	Mei. ♂, Ppd.	John and Henderson '62
Orthoptera	*Myrmeleotettix*	Mit. Mei. ♂, B Chrs.	John and Hewitt '65
Orthoptera	*Myrmeleotettix*	Mei. ♂, B Chrs., G. Bandg.	Gallagher *et al.* '73
Orthoptera	*Melanoplus*	Mei. ♂, Ab. C. Orient.	Henderson *et al.* '70
(4) Dermaptera	*Forficula*	Sex Chrs.	Callan '41

TABLE IX—*continued*

Group	Genus, etc.	Subject	Author
(5) Heteroptera	*Cimex*	Mei. ♂	Darlington '39
Heteroptera	*Pentatomidae*	Mei. ♂	Schrader '60
(6) Homoptera	*Aphis*	Parth.	Lawson '36
Homoptera	*Aspidoproctus*	Mei. ♂, Non-loc. C.	Hughes–Schrader '55
(7) Trichoptera	various	Sex Chrs.	Klingstedt '31
Trichoptera	various	Mei. ♀	Suomalainen '65
(8) Lepidoptera	*Apterona*	Mei. Parth.	Narbel '46
Lepidoptera	*Cidaria*	Mei. ♂, ♀	Suomalainen '65
(9) Coleoptera	*Micromalthus*	Hap. ♂	Scott '36
Coleoptera	*Tenebrionidae*	Mit. Mei. ♂	S. G. Smith '52
Coleoptera	*Otiorrhynchus*	Parth. Ppd.	Suomalainen '40, '54
Coleoptera	*Chilocorus*	Mit., Mei. ♂, Hyb.	S. G. Smith '66
(10) Diptera	*Miastor*	Mei. ♀, Dim.	White '46
Diptera	*Sciara*	Mei. ♀, Emb.	Carson '46
Diptera	*Tipula*	Mei. ♂	B. John '57
Diptera	*Drosophila*	Mit. Mei. ♂	Demerec '50
Diptera	*Chironomus*	Polytene	Beermann '52
Diptera	*Phryne*	Polytene	B. E. Wolf '56
(11) Hymenoptera	*Habrobracon*	Hap. ♂	R. L. Anderson '36
Hymenoptera	*Diprion*	Mei. ♀ Parth.	S. G. Smith '40
7. CHORDATA			
Pisces	*Salmo*	Mei. (abn.) ♂	Svärdson '45
Pisces	*Mogrunda*	Mit. Mei. ♂	Nogusa '55
Amphibia			
Urodela	*Triturus*	Mit. Emb.	Fankhauser '37, '41
Urodela	*Triturus*	Mei. ♂ (Hyb.)	Callan *et al.* '51

ANIMALS—*Continued*

Reptilia	*Chameleon*	Mit. ♂, ♀	Matthey et al. '56
Aves	*Columba*	Mit. ♂, ♀ Mei. ♂	Makino et al. '56
Mammalia	*Potorous*	Mit. Mei. ♂	Sharman et al. '52
Mammalia	*Microtus*	Mit. (X Y)	Sachs '53
Mammalia	*Cricetus*	Mei. ♂	Koller '46
Mammalia	*Cricetus*	Mit. Ppd.	Sachs '52
Mammalia	*Rattus*	Mit. (cancer)	Tjio et al. '56
Mammalia	*Mus*	Mit. (ascites)	Levan et al. '53
Mammalia	*Mus*	Mei. ♂, ♀, G. Bandg.	Polani '72
Mammalia	*Acomys*	Mit., Mei. ♂ Hyb.	Wahrman and Goitein '72
Mammalia	*Homo*	Mit., Mei. ♂	Sasaki and Makino '65
Mammalia	*Homo*	Mit., Mei. ♂	Eberle '66
Mammalia	*Homo*	Mei. ♂ (Mongolism)	Hungerford et al. '70
Mammalia	*Homo*	Mei. ♀	Edwards and Fowler '70
Mammalia	*Homo*	Mit. (cancer)	Koller '46
Mammalia	*Homo*	Mit.	Tjio et al. '56

PLANTS

ALGAE

Rhodophyceae	*general*	—	Drew '55, Godward '48
Phaeophyceae	*Plumaria*	Mit. Ppd.	Drew '39
Chlorophyceae	*Halopteris*	Mit. Ppd.	Ernst-Schwarzenbach '57
Chlorophyceae	*Cladophora*	Mit., Sp.	Geitler '36
Cyanophyceae	*Spirogyra*	C. (diffuse)	Godward '54
Characeae	various	—	Spearing '37
	Nitella	Mit. (exp.)	Karling '28

TABLE IX—*continued*

Group	Genus, etc.	Subject	Author
FUNGI			
Ascomycetes	*Peziza*	Mit. Mei.	Wilson '37
Ascomycetes	*Neurospora*	Mit. Mei.	McClintock '45
BRYOPHYTA			
Musci	various	Mit. Mei.	Vaarama '50
Musci	*Pleuorzium*	Mei. C.	Vaarama '54
Musci	various	Mit.	Wylie '57
Musci	various	Mei.	K. R. Lewis '57
Hepaticae	various	Mit. Mei. Sex Chrs.	Lorbeer '34
PTERIDOPHYTA	general	Mit. Mei.	Manton '50
PTERIDOPHYTA	*Pteridium*	Mit. (hap.)	Partanen *et al.* '55
GYMNOSPERMAE			
Coniferae	*Picea*	Mei. ♂	E. Andersson '47
Cicadales	various	Mit.	Sax and Beal '34
ANGIOSPERMAE	*Melandrium*	Sex Chrs.	Westergaard '40
ANGIOSPERMAE	Orchidaceae	Pollen	Barber '42a
ANGIOSPERMAE	various	Heterochromatin	La Cour '51
ANGIOSPERMAE	*Ranunculus*	End. Parth.	Rutishauser '54
ANGIOSPERMAE	*Trillium*	End. B Chrs.	Rutishauser '56
ANGIOSPERMAE	*Tradescantia*	Pollen	La Cour '49
ANGIOSPERMAE	*Agropyron*	Emb. Parth.	Hair '56
ANGIOSPERMAE	*Poa*, etc.	B Chrs.	Bosemark '57
ANGIOSPERMAE	various	Endomitosis	Tschermak-Woess '56
ANGIOSPERMAE	*Bromus*	Hyb. mei.	Walters '54
ANGIOSPERMAE	*Luzula*	Mei., Non-loc. C.	Malheiros *et al.* '47

TABLE X – CHROMOSOME CALENDAR

	PLANTS		
ALL THE YEAR ROUND			
Vicia faba	RT	$n = 7$	Lewitsky '31
Ranunculus spp.	RT		Larter '32
Allium spp.	RT	$n = 7, 8$	Levan '31
Crepis spp.	RT	$n = 3$	Babcock and Navashin '30
Rhoeo discolor	$\left\{\begin{array}{l} \text{RT} \\ \text{PMC} \\ \text{PG} \end{array}\right.$	$n = 6$	Darlington '29
Nothoscordum fragrans	RT, PMC, PG	$n = 8, 9, 10$	Dyer unpub.
Tradescantia paludosa	PMC, PG	$n = 6$	Anderson and Sax '36
JANUARY			
Hyacinthus orientalis	RT	$\left.\begin{array}{l} 2x = 16 \\ 3x = 24 \end{array}\right\}$	Darlington *et al.* '51
Scilla non-scripta	RT, PMC	$n = 8$	Darlington '26
FEBRUARY			
Uvularia spp.	PMC, PG	$n = 7$	Barber '41
MARCH			
Fritillaria meleagris	PMC	$n = 12$	Darlington '35, '36
Trillium cernuum	PMC	$n = 5$	Dyer unpub.

TABLE X—*continued*

PLANTS—continued

APRIL			
Paeonia spp.	PMC, PG	$n = 5, 10$	Dark '33, Stebbins *et al.* '39
Paris quadrifolia	PG	$n = 10$	Darlington '41
Trillium (until Sept.)	RT	$n = 10$	D. and La Cour '40
MAY			
Tradescantia spp. (till October)	⎱ PMC, PG, PT ⎰	$n = 6, 12$	Many authors
Chara and *Nitella*	filaments and antheridia	various	Karling '28
Allium spp.	⎱ RT, PMC, PG ⎰	$n = 7, 8, 9$ $(2x - 12x)$	Levan '31–'39
JUNE			
Secale spp.	PMC	$n = 7 (+ B)$	Müntzing *et al.* '41
Lilium spp. and hybrids	PMC, PT	$n = 12$	Richardson '36
Kniphofia spp.	PMC	$n = 5$	Darlington '33
Oxalis dispar (until Sept.)	PMC	$n = 6$	Marks '57
Campanula persicifolia	PMC	$n = 8, 16$	D. and La Cour '50
JULY			
Zea mays	PMC	$n = 10$	McClintock '33–'40
AUGUST			
Trillium grandiflorum	PMC	$n = 5$	La Cour unpub.
SEPTEMBER			
Tulipa gesneriana ($2x$)	PMC	$n = 12$	Upcott '37

OCTOBER			
Tulipa gesneriana (3x)	RT, PMC	$2n = 36$	Upcott '37
Hyacinthus, vars. —	PMC, PG	$2n = 16, 17, 19$, etc.	Darlington *et al.* '51
Narcissus pseudo-narcissus (2x, 3x, 4x)	RT, PMC	$2n = 14, 21, 28$	Nagao '33
Galanthus (until Feb.)	RT, PMC	$2n = 24, 48$	La Cour unpub., Sato '37
Crocus biflorus (until Feb.)	RT	$2n = 8$	Mather '32
NOVEMBER			
Narcissus poeticus	PMC, PG	$2n = 14, 21, 28$	Nagao '33
Hyacinthus vars. —	PG	$3x = 24$, etc.	Darlington '26
Hyacinthus var. Blue Giant	PG	$4x = 32$	Fogwill unpub.
DECEMBER			
Trillium erectum	PMC	$2x = 10$	Huskins *et al.* '35
Fritillaria pudica	PMC, PG	$\left.\begin{array}{l} 2x = 26 \\ 3x = 39 \end{array}\right\}$	Darlington '36
Narcissus bulbocodium	PMC and PG	$2x = 14$	Nagao, Wylie '52
N. biflorus	PMC and PG	$7^{II} + 10^{I}$	
Aucuba japonica	PMC	$4x = 24$	Meurman '29
Taxus baccata ♂	PMC	$2x = 24$	Maude '39
Helleborus foetidus	PMC	$4x = 32$	Maude '39

TABLE X—*continued*

NOTES

Plants show the following sequence of stages of division: 1, Pollen mother cells (PMC); 2, embryo sac mother cells (EMC); 3, Pollen grain (PG) first mitosis; 4, Endosperm.

PMC divisions (meiosis) take place when the anther is still translucent and about one-third of the length at maturity.

PG division may take place at any time, usually about half-way, between meiosis and maturity, when the pollen is becoming dry and the anther is turning yellow (or the proper colour of ripeness). Endosperm in suitable plants can be observed 3–20 days after fertilisation.

For other conditions see Ch. 4E.

ANIMALS

ALL THE YEAR ROUND		
Amphibia: tadpole's tails (if starved)	mitosis	
Mammalia: rat or mouse testis	mitosis and meiosis	{ Prokofieva '33, Barber and Callan '41
Diptera: *Drosophila* spp. salivary glands	polytene	Koller and Darlington '34
		cf. Chap. 8
Male locusts in culture, testes, *e.g.*		
Schistocerca gregaria		
4th larval instar	mitosis	John and Henderson, '62
Adult	meiosis	
Periplaneta (male)	meiosis } in interchange heterozygotes	John and Lewis, '57
APRIL–SEPTEMBER		
Culex, Chironomus, testes	meiosis }	*cf.* Ch. 8
Chironomus, salivary gland	polytene	
MAY–SEPTEMBER		
Orthopteran testes	mitosis and meiosis	{ John and Lewis '57, White '34–'40
JUNE–JULY		
Amphibian testes	mitosis and meiosis	Callan '42, Wickbom '45
JULY–SEPTEMBER		
Male grasshoppers, *e.g.*		
Chorthippus parallelus and *C. brunneus*		
1st larval instar	mitosis	John, Lewis and Henderson, '60
Adult	meiosis	
AUGUST		
Oligochaetan eggs and testes	meiosis	Muldal '52

APPENDIX II

Standard Solutions

(a) FIXING SOLUTIONS

TABLE XI – PROPERTIES OF REAGENTS

Reagent	Formula	M.W.	M.P. °C	B.P. °C	Saturated Soln. at 15° C	O or R	S.T.
1 Ethyl alcohol 'absolute'	C_2H_5OH	46	—	78°	Aqu. all propns.	—	22·3
2 Acetic acid 'glacial'	CH_3COOH	60	17°	—	Aqu. all propns.	—	27·6
3 Formaldehyde (Aqu. formalin)	$HCHO$	30	—	-20°	Aqu. 35–40%	R	—
4 Chloroform	$CHCL_3$	126·5	—	61°	Al. all propns.	—	27·1
5 Chromium trioxide (chromic acid)	CrO_3	100·01	196°	—	Aqu. 62%	O	—

TABLE XI—*continued*

Reagent	Formula	M.W.	M.P. °C	B.P. °C	Saturated Soln. at 15° C	O or R	S.T.
6 Osmium tetroxide (osmic acid)	OsO_4	255	40°	—	Aqu. 6% g.p. 100 ml.	O	—
7 Potassium dichromate	$K_2Cr_2O_7$	294	236°	—	Aqu. 9%	O	—
8 Picric acid	$C_6H_2(NO_2)_3OH$	229	122°	—	Aqu. 1·4%	O	—
9 Mercuric chloride (corrosive sublimate)	$HgCl_2$	272	275°	—	Aqu. 6% Al. 30%	(O)	—

NOTES ON TABLE XI

1 O or R: Oxidising or reducing agent. Formaldehyde must not be mixed with chromic or osmic acids until immediately before use. Osmic acid solution on the other hand is unstable in light except in the presence of chromic acid or mercuric chloride.

2 Aqu. = Aqueous. Al. = alcoholic.

3 S.T.: Surface tension, γ, in dynes per cm² at 20° C; γ for water is 72·8, for its mixtures with 1 and 2 it is increased roughly proportionately (from Hodgman's *Handbook of Chemistry and Physics*, 22nd edn).

4 Mercuric chloride is poisonous; osmic acid is also dangerous to the cornea.

5 Absolute alcohol is used only where water is to be excluded. Otherwise rectified spirit, 96%, can take its place.

TABLE XII – COMPOSITION OF COMPOUND FIXATIVES

(Aqueous solutions by volume unless otherwise stated)

(1)   *Osmic Fixatives*

Stock solutions (Material)	Flemming 1882			La Cour 1931			Minouchi	Champy 1913 Koller
	Strong Bulk	Medium Smear	Benda Chondrios	2BD General	2BE Plant	2BX Bulk	Animal	Mammal
Chromic acid 2%	100	100	100	100	100	100	100	100
Pot. dichromate 2%	–	–	–	100	100	100	300	–
Osmic acid 2%	53	66	66	60	32	120	120	50
Acetic acid 10%	133	80	15	30	12	60	–	–
Saponin 1%	–	–	–	20	10	10	–	–
Water, distilled	–	190	120	210	90	50	–	125
Totals	286	436	301	520	344	440	520	275

TABLE XII—*continued*

(2)   Formalin Fixatives

| Solutions | S. Navashin 1910 | | | | Bouin 1896 |
	M. Navashin 1925 (RT)	Sanfelice White 1940 (Animal)	Randolph 1935 (Plant)	Karpechenko 1927 (PMC)	Allen's B.15 (Animal)
Chromic acid 2%	100	100	100	100	1·5 g
Acetic acid 20%	100	60	70	67	5·0 g
Picric acid	—	—	—	—	1·0 g
Urea	—	—	—	—	2·0 g
Water, distilled	20	60	170	300	75 ml
Formaldehyde 40%	80	100	60	11	25 ml
Totals	300	320	400	478	100 ml

(3)   Alcoholic Fixatives

Components	Acetic alcohol	Carnoy 1886	La Cour 1944*	Smith 1940	Carnoy-Lebrun
Alcohol, abs.	100	100	100†	100	100
Acetic acid, glac.	33	16	2	7	100
Chloroform, pure	—	50	—	—	100
Formaldehyde 40%	—	—	7	40	—
Corr. sublimate	—	—	—	—	saturation
Water	—	—	70	—	—
Totals	133	166	179	147	300

* Blood smears.   † Methyl instead of Ethyl Alcohol.

(*b*) STAINS FOR CHROMOSOMES

1 *Aceto-carmine* (after Belling 1926, 1928; *cf*. Lima-de-Faria and Bose 1954).
   45 ml glacial acetic acid.
   55 ml distilled water.
Add 0·5 g of carmine, boil gently for 5 min in a reflux condenser.
Shake well and filter when cool.
A drop or two of 45% acetic acid saturated with iron acetate can be added. Too much iron precipitates the carmine. Stronger concentrations of carmine can be used if desired.

2 *Acetic-orcein* (after La Cour 1941).
The standard solution is 1% in 45% acetic acid. American synthetic orceins need greater dye concentration. Because of deterioration in dilute acid the stain is best kept as a 2·2% stock solution in glacial acetic acid which can then be diluted to 45% as required.
Thus dissolve 2·2 g orcein in 100 ml glacial acetic acid by gentle boiling. Cool, dilute, and filter as required.

3 *Acetic-lacmoid* (La Cour, 1st edn).
Solubility differs with dye source. The stain, which as an indicator is known as lacmoid, as a dye is known as resorcin blue.
Standard solution 1% in 45% acetic acid, prepared from stock, as for orcein.

4 *Leuco-basic fuchsin* (modified formula after De Tomasi 1936 and Coleman 1938).
   Dissolve 1 g basic fuchsin by pouring over it 200 ml boiling distilled water.
   Shake well and cool to 50° C.
   Filter: Add 30 ml N HCl to filtrate.
   Add 3 g $K_2S_2O_5$.
   Allow solution to bleach for 24 hours in a tight-stoppered bottle, in the dark; add 0·5 g decolorising carbon.*
   Shake well for about a minute and filter rapidly through coarse filter paper. Store in a tightly-stoppered bottle in the dark, at 4° C.

4′ *SO_2 water*.
   5 ml N HCl.
   5 ml $K_2S_2O_5$ 10%.
   100 ml distilled water.

4″ *Desoxyribonuclease (D N-ase) reagent*, Schedule 15.
Activation with magnesium is essential for the action of D N-ase on DNA. Dissolve 3·6 mg of $MgSO_4$ in 10 ml of glass-distilled water adjusted to pH 6 by adding 0·1 M NaOH.
   Use: 0·25 mg–0·5 mg D N-ase per 1 ml of $MgSO_4$ solution.
   Note. Use only crystalline D N-ase from a reliable source. It is inactivated by heat and is best stored at 4° C. Some samples of D N-ase dissolve slowly; be sure it is all dissolved before using the solution. Freshly prepared solutions of D N-ase can be stored in a deep-freeze cabinet at −20° C for at least 35 days without appreciable loss of activity. Gelatin (0·1%) is sometimes added as a stabiliser.

   * *Norit*, a proprietary vegetable carbon, is recommended.

5   *Crystal violet* (after Newton).
  1   1% aqueous solution boiled and filtered.
Also used as 0·5% and 0·1% solutions.
  2   1% iodine, 1% potassium iodide in 80% alcohol.

6   *Heidenhain's haematoxylin*.
  0·5% aqueous solution, ripened for 1–2 months, or rapidly by Hance's method (1933).

7   *Methyl-green-pyronin* (Unna Pappenheim, after Kürnick 1955, Schedule 14).
    Solution A: pyronin Y 2% aqueous solution.
    Solution B: methyl green 2% aqueous solution.
Before use, extract impurities from both solutions by shaking with equal volumes of $CHCl_3$. At least four successive extractions with fresh $CHCl_3$ are usually required.
    For use: 5 parts A, 3 parts B.

8   *Orange G–aniline blue* (Mallory, after La Cour *et al.* 1958, Schedule 12).
    2 g orange G.
    0·5 g aniline blue.
    100 ml pot. citrate buffer pH 3.

9   *Reaction mixture for arginine test* (Schedule 16).
    30 ml NaOH 1%.
    0·6 ml 2, 4 dichloro-α-naphthol 1% in 70% alcohol.
    1·2 ml sodium hypochlorite 1%.
Mix immediately before use. The 1% sodium hypochlorite can be obtained by dilution of a commercial solution of known concentration which is stored at 4° C.

10   *Buffered thionin* (Schedule 17).
    A. Saturated solution of thionin in 50% alcohol (filtered).
    B. Michaelis buffer:
    9·714 g sodium acetate ($3H_2O$)
    14·714 g sodium barbiturate
    500 ml distilled water (CO_2-free)
    C. 0·1 N HCl
    To make staining solution, mix 4 ml of A, 28 ml of B, 32 ml of C. The pH should be 5·7 ±0·2.

11   *Carbol fuchsin* (Schedule 20).
    A. Stock stain
    10 ml of 3% basic fuchsin in 70% alcohol
    90 ml of 5% phenol in distilled water.
    B. Staining solution
    45 ml of Solution A
    6 ml glacial acetic acid
    6 ml formalin.

12 *Combined stain-mounting medium*

Dissolve 70 g dimethyl hydantoin formaldehyde resin in 30 ml distilled water. The resin will take about a week to dissolve, if it is first broken into small lumps with a hammer and the solution shaken periodically.

The solution is then filtered and a 2% aqueous solution of toluidine blue is added until trial shows that a satisfactory staining intensity is reached.

(*c*) STAINS FOR STYLES

1 *Acid Fuchsin—Light Green*.
   54 ml lactic acid.
   6 ml glycerine.
   1 ml acid fuchsin 1%.
   1 ml light green 1%.

2 *Cotton Blue—Lacto-Phenol*.

Equal Parts { lactic acid. / phenol. / glycerine. / water. }

   0·08% to 1% cotton blue dissolved in the above.

3 *Lacmoid-Martius Yellow*.
   0·005 g lacmoid (resorcin blue).
   0·005 g Martius yellow.
   10–15 ml water.
Adjust to pH 8 by the addition of a few drops of 1% ammonia.

4 *Delafield's haematoxylin*.

To 100 ml of a saturated solution of ammonium alum (containing about 10 g) add, slowly, a solution of 1 g of haematoxylin dissolved in 6 ml absolute alcohol. Expose to air and light for 1 week. Filter. Add 25 ml glycerine and 25 ml methyl alcohol.

Allow to stand (uncorked) until the colour darkens.
Filter and keep in a tight-stoppered bottle.
Allow the solution to ripen before use for a month.

(*d*) VARIOUS FLUIDS

1 *Physiological saline solution* (for salivary glands).
   0·67% sodium chloride.

2 *Ringer solution A* (for cold-blooded animals).
   0·65 g sodium chloride.
   0·025 g potassium chloride.
   0·03 g calcium chloride.
   (0·02 g sodium bicarbonate, approximately, to give a pH 7·0–7·4.)
   100 ml distilled water.

2′   *Ringer solution B* (for warm-blooded animals).
   0·85 g sodium chloride.
   0·025 g potassium chloride.
   0·03 g calcium chloride.
   100 ml distilled water.

3   *Cleaning Slides and Cover Slips.*
   (1) Mixture for separating slides and cover slips (Carlson 1935).
       1 part *n*-butyl alcohol.
       9 parts xylol.
   (2) To clean:
       10 g potassium dichromate.
       10 g sulphuric acid.
       80–120 ml water.
   (3) Wash in running water.
   (4) Store in 70% alcohol.

4   *Stain-fixing mounting media* (after Zirkle 1940).
   No. 1
       10 g gelatin.
       10 ml sorbitol.
       50 ml glacial acetic acid.
       60 ml distilled water.
       0·5 g $Fe(NO_3)_3 \cdot 9H_2O$.
       Carmine to saturation.
   Orcein, 1 g or lacmoid, 1 g can be substituted for the carmine. The iron salt
is then unnecessary.
   No. 2
       20 ml venetian turpentine.
       55 ml phenol (as loose crystals or 88% liquid).
       35 ml propionic acid.
       10 or 15 ml glacial acetic acid.
       25 ml distilled water.
       0·5 g $Fe(NO_3)_3 \cdot 9H_2O$.
       Carmine to saturation.
   Orcein or lacmoid can be used as above instead of carmine.
   Mix in a graduated cylinder in the following order: propionic acid, turpen-
tine, mix thoroughly; phenol, acetic acid, water. The ferric nitrate should be
dissolved before any carmine is added. Filter after 12 hours.
   At no time should the mixture be heated.
   Do not allow fluid to come into contact with the skin.

5   *Sealing medium for semi-permanent acetic preparations.*
   Dissolve 10 g gelatin (powder) in 100 ml of 50% acetic acid. Allow gelatin
to swell, then heat to dissolve. Apply with thin glass rod. Can be removed
by immersion in 45% acetic acid as is required in making permanent.

6   *Mounting medium for unstained stripping-film autoradiographs* (*permanent
mounts for phase-contrast*)
       3 g gelatin.
       0·2 g chrome alum.

80 ml distilled water.
20 ml glycerine.
Store at 4° C. To use: melt at 50° C and apply one or two drops to slide after moistening film, press cover slip down gently and let cool.

7 *Mayer's albumen slide fixative.*
    25 ml albumen (white of egg).
    25 ml glycerine.
    0·5 g sodium salicylate.
Filter before use.

8 *Haupt's gelatin adhesive.*
    1 g gelatin.
    100 ml distilled water.
Dissolve at 30° C, then add 2 g phenol crystals. Stir well and filter.
To use, place a small drop on a clean slide and film with a finger tip. For squashes of plant tissues harden the film by dipping the filmed slide briefly in 95% alcohol, for paraffin sections use 10% formalin.

9 *Bleaching solution.*
    1 part hydrogen peroxide (20 vol.).
    3 parts alcohol 80%.
To be used fresh.

(*e*) DROSOPHILA MEDIUM

Composition of Drosophila *Food* (Bridge's formula, for salivary glands. 1932. *cf. Dros. Inf. Service*, p. 62, 1936).
    Water, 75 ml.
    Treacle, 13·5 ml.
    Maize meal, 10 g.
    Agar-agar, 1·5 g.
    Nipagin or Moldex, trace.
This quantity is enough for three half-pint milk bottles. Fill 1 in deep. The nipagin preserves the food from mould.
Soak the agar 12 hours in water, boil it until it dissolves, add treacle and cornmeal and stir.
When cool a small portion of food should be removed from one side and a piece of folded, sterilised crepe paper inserted for the larvae to pupate on.
The medium should be yeasted by the addition of a few drops of yeast suspended in water (of a creamy consistency). Dried yeast is liable to become mouldy. Flies are mated for two days in 3″ × 1″ vials with a small wedge of food at about 25° C before moving to the bottles prepared as above. All bottles are stoppered with cotton wool.
D. melanogaster, as well as most of the larger species, can be kept at room temperature.
Fattening larvae for salivary glands. To avoid overcrowding of larvae, from the laying of too many eggs, transfer parents after a day. As soon as the young larvae appear (in *melanogaster*, after about three days) the cultures should have a second yeasting. They should then be kept at about 18° C until ready for the knife, to slow down growth. To do so, bottles can be placed in trays which are continuously flooded with tap water.

(f) PHOTOGRAPHIC SOLUTIONS

1   *Developers.*
    (i) Except for special purposes it is generally best to follow the makers' recommendations.
(ii) *Kodak, D 19 High-Contrast Developer, for Stripping-Film Autoradiography.* (AR 10 plates).
    2 g 'Elon'.
    90 g sodium sulphite (anhyd.).
    8 g hydroquinone.
    45 g sodium carbonate (anhyd.).
    5 g potassium bromide.
    Water to make up 1000 ml.
  Filter before use.

2   *Fixing Bath (acid).*
    80 g sodium hyposulphite.
    5 g potassium metabisulphite.
    Water to make up to 250 ml.

3   *Reducer*
    Potassium ferricyanide, 10% solution.
    For use, add a few drops of the above to a weak non-acid hyposulphite solution. The colour is a fair indication of the strength of the reducer; it should be a pale yellow colour. Wash the negative thoroughly after reducing.

(g) TRANSFER SOLUTION FOR AUTORADIOGRAPHY STOCK SOLUTION

    200 g sucrose.
    0·1 g potassium bromide.
    Dissolve in water to make up 1000 ml.
  For use, dilute 1 vol. with 9 vols of water.

Schedules of Treatment

SCH. 1. PARAFFIN PREPARATION.

A. *Fixing:* 12–24 hours in 10 ml glass phial (aqueous fixation).
B. *Washing:* 3 quick changes of water in phial.
C. *Dehydration:* in water–alcohol mixtures by steps.
     50% 3 hours.
     70% overnight.
     80% 3 hours.
     95% 3 hours.
     100% overnight.
D. *Transference:* by steps in alcohol–chloroform mixtures.
     25%, 50%, 75%, 2 hours in each,
followed by pure chloroform to which wax has been added.
E. *Infiltration:* Place on oven top or hot plate at about 30° C. Add more wax at intervals for 2 days. Place in small watch-glass inside oven at 60° C, adding more wax, for 4 hours.
F. *Embedding:* Transfer contents of watch-glass with more molten wax to a paper boat or flat-bottomed watch-glass smeared inside with glycerine. Orientate and group material. Allow skin to form on wax. Submerge carefully in cold water.
G. Prepare block and cut sections on microtome.
H. Lay ribboned sections shiny face downwards on filmed slide covered with water or 20% alcohol.
K. Stretch and straighten ribbon by placing slide on hot plate for 5 min.
L. Drain water off slide and leave to dry on hot plate for 4–24 hours.
M. Xylol, 10–15 min to dissolve wax ribbon.
N. Absolute alcohol, 2 min.
O. Bleach in 20 vol. H_2O_2 and 80% alcohol, 1 : 3, 4–12 hours after osmic fixatives.
P. Pass quickly through alcohol series, 80%, 40%, to water.

SCH. 1'. RAPID DIOXAN METHOD FOR ROOT TIPS (La Cour 1937).

A. *Fixing:* 12–24 hours in 10 ml glass phial (aqueous fixation).
B. *Washing:* 3 quick changes of water in phial.
C. *Dehydration:* In water–dioxan mixtures by steps.
     25%, 50%, 75%, 2 hours in each. 100% dioxan overnight.
E. *Infiltration:* Place directly in oven, adding wax of low m.p. at intervals for 4 hours. Transfer to pure molten wax for 2 hours.

SCH. 1″. SOFTENING METHOD FOR ANIMAL EGGS (after S. G. Smith 1940).

A. *Fixation:* Kahle's Fluid without water, 2 hours (for Feulgen staining).
[B. No washing.]
C and D. *Dehydration:* By stages as follows:

		Water	Ethyl alcohol	n-Butyl Alc.	Phenol
(i)	1 hr	30%	50%	20%	—
(ii)	24 hr	11%	50%	35%	4%
(iii)	1 hr	5%	40%	55%	—
(iv)	1 hr	—	25%	75%	—

follow by two changes of 4% phenol in *n*-butyl alcohol.

E. *Infiltration:* Place directly in oven with equal amount of paraffin wax.
Change to fresh molten wax after about 16 hours.

F. *Embedding:* As usual. In preparing for sectioning cut down the block.
Expose one side of the material and soak the block for 24 hours or more in
water before cutting.

A softening medium consisting of 9 parts 60% ethanol, 1 part glycerin, is
recommended by Baker (1941).

SCH. 1A. ESTER WAX PREPARATION (after Steedman 1960).

A. After washing: leave tissues in 70% alcohol overnight.

B. Equal parts (by vol.) 70% alcohol and ethylene glycol monoethyl ether
('cellosolve').

C. Equal parts (by weight) 'cellosolve' and ester wax in oven at about
40° C overnight.

D. Molten ester wax (freshly melted at 45–47° C), 3 changes 30 min in each.

E. *Embedding:* Heat a small amount of ester wax at about 50–55° C: use
new wax which has not previously been melted. Pour the melted wax into a cold
embedding dish, previously smeared inside with glycerine. When a thin layer
of congealed wax has formed at base of dish, melt surface of wax with a hot
spatula, introduce specimen and orientate in desired position. Surround sides of
dish with ice-cold water; keep surface of wax melted until the block solidifies,
to avoid air creeping into the block: do not plunge dish below water.

F. *Sectioning:* Sections are best cut 1–2 days (or longer) after embedding.
Mount and trim block in the usual way; cut sections at 30–50 per min. Stretch
and straighten ribbons on water on the slide at about 30–35° C; drain off water
as soon as sections regain their original size. Dry at 30–40° C overnight. Remove
wax in xylol.

SCH. 1A′. Alternatively.

A. After washing, leave in 70% alcohol 3 hours, 95% alcohol overnight.

B. Equal parts (by weight) 95% alcohol and ester wax in oven at about
40° C overnight.

C. Molten ester wax, 3 changes.

D, E and F as above.

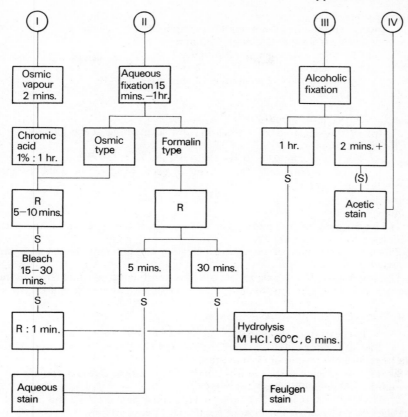

SCHEDULE 2.—Showing the general combinations of treatments possible in smear methods. R, rinse in running water. S, store if necessary in 70% alcohol up to 2 months. (S) store only up to 48 hours. I, for animals and bacteria only. IV, stain-fixative method. III–IV, compare Sch. 2'. *Note:* Hydrolysis for Feulgen staining, which takes 6 minutes after alcohol fixation, must be extended to 10 minutes after aqueous or vapour fixation. In certain circumstances it may be necessary to extend any of these times, *e.g.* after long storage, for some species and for some tissues.

SCH. 2'. PERMANENT STAIN-FIXATION.

Tissues: SMC, PMC, PG, PT, ganglia.

A. Dissect out small pieces of tissue from acetic alcohol and crush them in a drop of stain-fixative, on a very clean slide. The blunt end of a bone or aluminium needle holder is a suitable tool. Leave 1–2 min.

B. Prepare cover slip by smearing thinly with Mayer's albumen and drying over spirit flame 1–3 seconds.

C. Remove all but the smallest debris and place cover slip in position.

D. Pass the slide quickly over a spirit flame 5–6 times. The solution must not boil; judge the heat by passing the slide over the palm of the hand.

After aceto-carmine:

E'. Invert the slide in a smearing dish containing 10% acetic acid. (The cover slip will separate from the slide after 5–15 min.)

F'. Take the slide and cover slip through:

      1 in 3 acetic-alcohol   2 min

      Absolute alcohol, 2 changes   2 min each.

Mount in 'Euparal'.

After acetic-orcein or lacmoid:

E". Invert slide in a covered smearing dish containing acetic alcohol (1 : 3).

F". Pass through:

      Absolute alcohol, 2 changes 1 min in each.

      Mount in 'Euparal' after orcein, in cedarwood oil or aged 'Euparal' after lacmoid.

Notes:

(1) Use no more stain-fixative than will permit of the cover slip being placed in position without air bubbles.

(2) Vary the amount of intermittent heating for different tissues and stages of division.

(3) Mount the cover slip rapidly to avoid absorption of moisture. If cloudiness appears, place the slide on the hot plate for a short time and it will clear.

SCH. 3. ACETIC-LACMOID SQUASH METHOD.

      Tissues: Root tips, embryo sacs, pollen grains.

[A. Fix in acetic alcohol 12–24 hours.] For rapid counts in root tips, fixation can be reduced to 10 min or even omitted.

B. Stain by placing tissues in a watch-glass containing a few drops of: 10 ml standard acetic-lacmoid, plus 1 ml N HCl.

C. Heat without boiling 2–3 times over a spirit flame. (Vary the amount of heating according to the hardness of the tissues.) Leave 10 min.

D. Tease out the tissues on the slide in a drop of fresh standard solution.

E. Place filmed cover slip in position. Apply pressure under several thicknesses of blotting paper, allowing no sideways movement of the cover slip.

F. Mount as in Sch. 2'.

Note. Lacmoid can be replaced with orcein if desired (*cf*. Tjio and Levan 1954).

SCH. 3A. RAPID TUMOUR SQUASHES (Koller 1942).

A. Fix small pieces in acetic alcohol, 10 min–24 hours.

B. Transfer shreds of tissue. 10% acetic acid 5 min, 45% 10 min.

C. Stain in acetic-lacmoid or -orcein at 40° C for 15–30 min.

D. Macerate stained tissue in a drop of the stain on slide.

E. Remove unmacerated fragments.

F. Cover with filmed cover slip and press to spread cells.

G. Seal, or make permanent as in Sch. 2.

SCH. 4. ACETIC-ORCEIN SQUASH METHOD.

      Tissue: Salivary glands, ganglia, *Drosophila*.

The best stain in general is: 1% orcein in 45% acetic acid.

For *Drosophila melanogaster* use 2% stain in 70% acetic acid.
A. Dissect out glands in saline solution or in acetic-orcein.
B. Leave glands in a drop of stain on a well-slide 5–10 min.
C. Film the cover slip.
D. Transfer to clean slide by fine pipette.
E. Flatten by applying slight pressure under blotting paper.
F. Seal, or make permanent as in Sch. 2′.

SCH. 4′. SCIARA.

The best stain is :. 10 ml of 1% orcein in 45% acetic acid plus 1 ml chloroform.
A. Dissect out glands in 45% acetic acid.
B. Stain 2–3 min.
Subsequent treatment as above.

SCH. 4″. CHIRONOMUS.

The best stain is : 2% orcein in 50% acetic acid.
A. Dissect out glands in saline solution.
B. Stain 3–5 min.
Subsequent treatment as above.

SCH. 4‴. LACTIC-ACETIC-ORCEIN METHOD.

Tissue: Salivary glands and animal chromosomes generally (Vosa 1961).
Dissolve 2 g of orcein (natural) in 100 ml of a solution consisting of equal parts lactic acid and glacial acetic acid. Filter, leave overnight and filter again. The working solution is obtained by diluting the stock to 50% with glass distilled water. Subsequent treatment as above.

SCH. 5. FEULGEN SQUASH METHOD (Heitz 1936, Darlington and La Cour 1938, Hillary 1940, Battaglia 1957).

Tissues: All, except endosperm.
A. Fixation: 4–24 hours (see Table XI).
B. Rinse, 2–3 changes of water, more after formalin.
C. Macerate by hydrolysis in N HCl at 60° C for 6 min or more as in Table XIII.

TABLE XIII – MACERATION METHODS FOR SCHEDULE 5

(H, hydrolysis. Range of times according to hardness of tissue)

Fixative	Plant	Animal
Alcoholic (hard tissues and Feulgen-photometry)	RT and ovaries 6 min H ——————————— PG M$_3$ and 6 min H	6 min H
Osmic	10–20 min H	10 min H
Formalin	M$_3$ and 10 min H	15 min H

D. Stain in leuco-basic fuchsin: animal 1–2 hours, plants 2–3 hours.

E. Tease out small pieces of tissue (root tips, thin slices of tip) on a slide with a drop of 45% acetic, with the blunt end of a bone needle holder, to obtain small groups of cells.

F. Film the cover slip, place it in position and apply pressure under several thicknesses of blotting paper, allowing no sideways movement of cover slip.

G. Heat the slide gently over a spirit flame 4 or 5 times, do not boil.

H. Separate the slide and cover slip by turning the slide face down in a smearing dish containing 40% alcohol; after 3–10 min the cover slip will fall off.

J. Pass the cover slip (and slide if necessary) through alcohols: 80% 2 min, absolute 2 changes, 2 min each.

K. Recombine slide and cover slip by mounting in 'Euparal'.

SCH. 5A. FEULGEN SQUASH METHOD FOR ENDOSPERM (Rutishauser and Hunziker 1950; *cf.* La Cour 1954).

A. Fix developing seeds in acetic alcohol 1–2 hours, harden in 95% alcohol overnight.

B. 70%, 50% and 30% alcohol 10 min in each.

C. Rinse in water 10 min.

D. Hydrolyse in N HCl at 60° C 8–12 min.

E. Stain in leuco-basic fuchsin 2 hours.

F. Tap-water, 2 changes; leave 10 min.

G. Dissect out endosperm on the slide in a drop of 45% acetic acid, under a dissecting microscope using tungsten needles pointed in molten $NaNO_2$.

H. Film the cover slip, place in position and apply gentle pressure.

J. To make permanent, proceed as in Sch. 5 or Sch. 8.

Note. Weak staining can be corrected by the addition of acetic-orcein after dissection at step G.

SCH. 6. FEULGEN METHOD FOR SECTIONS AND SMEARS.

A. Distilled water: rinse,

B. Hydrolyse in N HCl at 60° C (see Sch. 2, legend).

C. Stain in leuco-basic fuchsin.* Plant tissues 2–3 hours, animal 1–2 hours.

D. Fresh SO_2 water* in stoppered jars, 3 changes of 10 min each.

E. Distilled water: rinse.

F. Alcohol series, 20%, 60%, 80%; rinse in each.

G. Absolute alcohol, 2–3 min.

H. Mount in 'Euparal'.

Notes:

(1) Rinsing in running water 2–3 min after staining increases the intensity of the stain, but omit in Feulgen-photometry and in tests for DNA.

(2) Smears fixed in acetic-alcohol, can be pressed to remove diffuse stain and spread the cells, as in Sch. 5, E and F.

* See Formulae of Reagents.

SCH. 6A. FEULGEN-LIGHT-GREEN METHOD (Semmens and Bhaduri 1941).

Tissues: All.

Fixatives: without acetic acid, or an aqueous fixative with acetic acid if treated before hydrolysis in 1% chromic acid for 4–6 hours, and afterwards thoroughly rinsed in water and placed in 75% alcohol for 4 hours.

(1) SECTIONS (Follow Sch. 1, A–P)

A. Rinse in water, leave in 75% alcohol, 2–3 hours.

B. Distilled water: rinse.

C. Hydrolyse in N HCl at 60° C, 10 min.

D. Stain in leuco-basic fuchsin, 2 hours.

E. Rinse in SO_2 water 2 changes, 10 min each.

F. Rinse in distilled water, 50% alcohol then 70% alcohol.

G. Mordant in 80% alcohol saturated with Na_2CO_3, 1 hour.

H. Rinse in 80% and 95% alcohol, 1 second each.

I. Stain in a filtered saturated alcoholic solution of light green, to which is added 2–3 drops of pure anilin oil, 20–25 min.

J. Drain off excess dye, rinse in a saturated solution of Na_2CO_3 in 80% alcohol, 10 ml; 80% alcohol, 90 ml.

K. Differentiate in 95% alcohol till green remains only in the nucleoli.

L. Dehydrate in absolute alcohol 2–3 changes, alcohol-xylol (1 : 1), alcohol-xylol (1 : 3).

M. Xylol, mount in neutral balsam.

(2) SQUASHES

Follow procedure in Sch. 5. After staining, tease out cells on slide in 45% acetic acid. Apply filmed cover slip and press. Heat slide gently over a spirit flame. Separate slide and cover slip in 40% alcohol. Proceed as above from G.

(3) SMEARS

Follow procedure in Sch. 6, A–D; then as above from F.

SCH. 7. RAPID TOLUIDINE-BLUE SQUASH METHOD (Marks 1973).

(For plant chromosomes in meristems and other somatic tissues, either taken direct from the plant or first pre-treated with colchicine or α–bromonaphthalene etc.).

The stain is used as a 0·05% solution made up in McIlvaine citric acid–Na_2HPO_4 buffer at pH_4.

A. Fix and hydrolyse tissues 15 min in 5 N HCl at room temperature.

B. Rinse tissues in distilled water. After washing, if so desired, they can be stored in water at 4° C for up to twenty-four hours without deterioration.

C. Macerate tissue with needle in a drop of stain on a slide, until it is broken up into small well-stained fragments. Staining occurs rapidly.

D. Place cover slip in position and apply pressure above larger fragments with a needle holder, as well as gentle overall finger-tip pressure through two or more layers of filter paper. Ring with rubber solution if only temporary preparation is required.

E. To make permanent, prise off the cover slip from slide on dry ice (as in Schedule 8), rinse in distilled water, dry in air and mount directly in 'Euparal'. Weak staining can be corrected after the cover slip is removed on dry ice, by flooding the slide with stain and finally rinsing in water.

SCH. 8. QUICK-FREEZE METHOD FOR MAKING SQUASH PREPARATIONS PERMANENT (Conger and Fairchild 1953).

For use after Feulgen or acetic-orcein.

A. Lay back of slide flat on block of dry ice, pressed down with weight to insure good contact.

B. Freeze 30 seconds, longer does no harm.

C. Prise off cover slip by lifting at one corner with a razor blade, while the slide is still on the dry ice.

D. Transfer slide and/or cover slip immediately to 95% alcohol, before thawing. Leave 5 min.

E. Absolute alcohol 5 min, longer does no harm.

F. Mount in thick 'Euparal', leaving an excess of alcohol on slide. Leave 2 days to dry.

Note. The material mostly sticks to the slide if no adhesive is used and mostly to the cover slip if this is filmed.

SCH. 9. CRYSTAL VIOLET METHOD.

Material: Sections or smears after aqueous fixation.

A. Rinse.

B. Stain: $\begin{cases} 0.5\%, \text{ 3--10 min.} \\ 0.1\%, \text{ 10 min--1 hour.*} \end{cases}$

C. Rinse.

D. 1% I_2, 1% KI in 80% alcohol, 30--45 seconds.

E. 95% alcohol, rinse.

F. Absolute alcohol, 4--10 seconds.

G. Clove oil; differentiate under microscope about 30 seconds.

H. Xylol; 3 changes, 10 min.

K. Neutral balsam or 'Clarite', mount.

Notes:

(1) The quickest staining and differentiation are best.

(2) Fading is differential; the centromere is often the last structure to fade.

(3) Faded or under-stained slides can be re-stained after removing cover slip in xylol and taking down through alcohol to water. Metaphase chromosomes may then show an additional contraction.

SCH. 9'. CHROMIC MODIFICATION FOR DEEPER STAINING (La Cour 1937).

A. Rinse.

B. Stain: 10 min. in 0.5%

C. Water rinse.

> Absolute alcohol, 2 seconds.

D. I_2-KI in 80% alcohol, 2 min.

> Absolute alcohol, 2 seconds.

> Chromic acid 1% aqueous polution, 15 seconds.

> Absolute alcohol, 5 seconds.

* White, unpub., for orthopteran testes.

Chromic acid 1% aqueous solution, 15 seconds.
E/F. Absolute alcohol, 10–15 seconds.
G/K. As before.
Note. This is the only crystal violet method that gives passable results after alcoholic fixatives.

SCH. 10. RAPID HAEMATOXYLIN METHOD.

Material: any smears or sections.
A. Rinse.
B. Mordant in 4% alum, 10–20 min.
C. Rinse in running water, 10–15 min.
D. Stain in 0·5% haematoxylin (ripened), 5–15 min.
E. Rinse in water and de-stain 5–20 min in saturated aqueous picric acid (Tuan 1930).
F. Blue the stain in a jar of water containing 1 or 2 drops of 0·880 ammonia, 1 min.
G. Rinse in running water, 30 min.
H. Pass quickly through an alcohol series, 20%, 60%, 80%, absolute.
I. Examine in clove oil.
J. Xylol, mount in balsam or 'Clarite'.

SCH. 11. GIEMSA-GELEI METHOD (Gelei 1921).

Material: animal or bacterial, best after osmic vapour—chromic acid smears.
Stain: { 10 drops Giemsa solution. / 10 ml water.
A. Mordant in 1–2% ammonium molybdate, 5–15 min.
B. Rinse in distilled water, 2–5 min.
C. Stain: 10–45 min.
D. De-stain in 96% alcohol.

SCH. 12. ORANGE G-ANILINE BLUE METHOD.

(For differential staining of heterochromatin at interphase and telophase, and of chromosomes at various stages of mitosis and meiosis. La Cour *et al*. 1958.)
Tissues: All, on sections only.
Fixative: Lewitsky's fluid (equal vol. 1% chromic acid: 10% formalin).
A. Rinse sections in pot. citrate buffer pH 3.
B. Stain, 3 min (maximum).
C. Rinse in pot. citrate buffer pH 3.
D. Drain and partially dry by blotting.
E. Dehydrate in tertiary butyl alcohol, two changes; 2–3 min in each.
F. Mount in 'Euparal'.

SCH. 13. AZURE B METHOD.

(For differential staining of sex chromosomes at prophase of meiosis, Saez 1952; *cf*. Flax and Himes 1952.)

Fixatives: Carnoy or acetic alcohol.
Material: Smears or sections of testes.
A. Rinse in distilled water.
B. Treat in distilled water at 50° C, 15 min.
C. Stain in Azure B (0·1–0·2 mg per 1 ml of citrate buffer pH 4 at 40° C, 3 hours).
D. Drain and partially dry by blotting.
E. Dehydrate in tertiary butyl alcohol, 2 changes, 2–3 min in each.
F. Mount in 'Euparal'.

SCH. 14. METHYL-GREEN-PYRONIN METHOD.

(Unna Pappenheim modified for RNA discrimination Brachet 1940a, 1944; Painter 1943; Davidson and Waymouth 1944; Kaufmann *et al.* 1951; Jacobson and Webb 1952.)
    Tissues: All.
Fixatives: Carnoy or acetic alcohol.
Smears or sections, use two slides or two halves of one slide for test and control.
A. Rinse in distilled water.
B. Apply R N-ase* to test slide; keep at 37° C, 1–2 hours.
C. Stain the test and the control together in methyl-green-pyronin, 10–30 min.
D. Rinse in distilled water.
E. Drain and partially dry by blotting.
F. Dehydrate in tertiary butyl alcohol, 2 changes, 5 min in each.
G. Mount in 'Euparal'.

SCH. 15. METHOD FOR USING D N-ASE.

(For removal of DNA or discrimination in conjunction with Feulgen method, Kunitz 1948, 1950; Kaufmann *et al.* 1951; Jacobson and Webb 1952; Ghosh 1955; Rasch and Woodard 1959.)
    Tissues: All.
Fixatives: Carnoy or acetic alcohol.
Smears of sections: use two slides or two halves of one slide for test and control.
A. Rinse in distilled water.
B. Apply solution of D N-ase† to test slide and $MgSO_4$ solution alone to control slide; keep both at 37° C, 1–2 hours.
C. Treat test and control together as in Feulgen method Sch. 6.
Note. 10% formalin may be used as a fixative if sections, smears or squashes are steeped in water at 90° C for 10 min prior to treatment with D N-ase.

* Crystalline R N-ase from a reliable source should be used. Methods of preparation have been described by Kunitz (1940) and McDonald (1948). Store at 4° C. For use, 0·5–1 mg is dissolved in 1 ml of glass-distilled water adjusted to pH 6 by adding 0·1 M NaOH. The control slide should be kept in the water alone at pH 6 for the same period of time as the treated slide. If two halves of one slide are used as test and control, the two halves can be separated by applying paraffin wax with a heated wire to form narrow-walled compartments of sufficient depth for treatment.
    † See Formulae of Reagents.

SCH. 16. MODIFIED SAKAGUCHI REACTION FOR ARGININE (McLeish *et al*. 1957, McLeish and Sherratt 1958).

Fixatives: 4% neutral formalin or Lewitsky's fluid (equal vol. 1% chromic acid : 10% formalin).

Material: Smears of isolated nuclei, sections and smears.

A. From water, pass through alcohol series.

B. Coat smears or sections with celloidin (to prevent loss) by dipping the slides in 1% celloidin dissolved in equal parts ether and absolute alcohol. Dry in air.

C. Transfer to reaction reagent,* 6 min.

D. Rinse rapidly 5% urea.

E. Transfer to 1% NaOH, 5 min.

F. Mount in a mixture of 9 vol. glycerine : 1 vol. 10% NaOH.

Note. The 4% neutral formalin is the most satisfactory fixative to use for quantitative studies. It is obtained by dilution of 40% formalin stored over marble chips.

SCH. 17. BUFFERED THIONIN METHOD.

(For differential staining of sex chromatin in human resting somatic nuclei, Klinger and Ludwig 1957.)

Fixatives: 95% alcohol, 10% formalin or Klinger and Ludwig's fixative.

Material: Smears or sections.

A. Coat smears or sections with celloidin film.

B. Rinse in distilled water.

C. Hydrolyse at 20–25° C in 5 N HCl 20 min.

D. Rinse thoroughly in several changes of distilled water.

E. Stain in buffered thionin† 15–60 min depending on specimen, using shortest time consistent with adequate staining.

F. Rinse in distilled water, then in 50% alcohol.

G. Rinse in 70% alcohol until clouds of stain cease to appear.

H. Dehydrate in 80%, 95% and absolute alcohol.

I. Xylol, mount in 'De Pex'.

SCH. 18. GIEMSA BANDING TECHNIQUE FOR RECOGNITION OF HETEROCHROMATIN.

(Pardue and Gall 1970, Arrighi and Hsu 1971, Hsu and Arrighi 1971, Sumner 1972, Vosa and Marchi 1972, Schweizer 1973, Vosa 1973.)

Material: Squashes of plant somatic tissues, animal somatic chromosomes generally in spread free cells; meiotic chromosomes in spread mammalian spermatocytes (Schedule 26). Giemsa stock solution (improved R 66, G. T. Gurr) is widely employed; use diluted, × 50 with M/15 Sörenson phosphate buffer pH 6·9.

A. Pre-treat tissues or free cells according to usual practice *viz*. plant cells usually with colchicine, animal cells with colcemid and/or hypotonic solution.

B. Fix either in acetic-ethyl alcohol or acetic-methyl alcohol as 1 : 3 solutions.

C′. To assist cell separation of plant tissues treat in 45% acetic acid at 60° C for 15–30 min, or place in 9 parts 45% acetic acid : 1 part N HCl in a watch-glass and heat gently over a spirit flame until solution steams; leave in solution on bench for 1 min. Macerate the tissues on a slide in a drop of 45%

* See Formulae of Reagents.

acetic acid, cover with a cover slip and pass slide once or twice over a spirit flame before applying gentle pressure from above.

C″. Spread free animal cells on slides by drying in air; leave 16 hours or longer before proceeding to the next step.

C‴. Separate cover slips from slides of squash preparations by Quick-Freeze Method (Schedule 8); dry slides bearing the cells in air, after removal of cover slip. Leave 1–2 days before proceeding further.

D′. Place slides bearing spread animal cells in 0·2 N HCl for one hour at room temperature (20° C) and then rinse in distilled or deionised water. Omit this step for plant cells.

D″. Steep slides in a coplin jar of freshly prepared 5% aqueous solution of barium hydroxide (Ba(OH)$_2$ 8H$_2$O) at 50–55° C (or a saturated solution at room temperature) for 5–20 min.

E. Rinse thoroughly with several changes of distilled or deionised water to remove scum that has formed.

F. Incubate preparations in 2 × SSC (0·3 M sodium chloride containing 0·03 M tri-sodium citrate, adjusted to pH 7 with 0·1 M citric acid) at 60° C for 30–120 min.

G. Rinse preparations briefly in distilled or deionised water and stain in diluted Giemsa stock solution for 1½–16 hours.

H. Rinse briefly in distilled or deionised water, blot carefully and dry in air.

I. Rinse briefly in xylene and mount in 'De Pex' or similar mountant.

Note. The use of 5% aqueous solution of barium hydroxide at 50–55° C was suggested by Sumner, Evans and Buckland (1971), as a modification of the original technique for centromeric heterochromatin in chromosomes of mouse and man. Sumner (1972) then found that this step can be omitted for other characteristic G-banding in chromosomes of various mammals. Barium hydroxide treatment, preferably with a saturated solution at room temperature (Vosa 1973) is, however, necessary for demonstration of such banding in plant chromosomes.

In plants, because of masking by stained cell walls, the best banded preparations are obtained when metaphase plates are extruded entire from the cells.

SCH. 18A. THE USE OF TRYPSIN FOR MAPPING G-BANDS (Seabright 1972: *cf.* Wang and Fedoroff 1972, Dutrillaux 1973).

Material: Human chromosomes in leucocytes of whole blood.
Culture procedure and cytological preparation:
Proceed as in Schedule 24 up to step H.
Digestion and staining:

I. To a phial of Bactotrypsin* add 10 ml of sterile isotonic saline and use 1 ml of this solution diluted with 9 ml of isotonic saline to flood the air-dried cells on slides.

J. Leave for 10–15 seconds in solution and then rinse twice with saline. The preparation can be examined while still wet, by phase contrast microscopy to assess the effect of the enzyme and, if necessary, can then be treated again for a further short period, until the chromosomes appear slightly swollen.

K. After examination, rinse again with saline and place immediately in Leishmann stain diluted 1 : 4 with Sörenson phosphate buffer pH 6·8 for 3–5 min.

* Difco Cat. no. 0153–59.

L. Rinse rapidly in buffer, blot carefully and leave to dry in air before mounting in a neutral mountant.

Notes:

(1) Some other proteolytic enzymes such as ficin, bromelin, papain, pronase and protease are apparently also capable of producing G-bands, when used in buffer at a pH giving maximal activity.

(2) Giemsa can be used to stain the bands in place of Leishmann. Use 1 part stock (Gurr's R66) in 10 parts Sörenson's phosphate buffer pH 6·8 and stain for 1–2 min.

SCH. 19. QUINACRINE FLUORESCENCE METHOD FOR RECOGNITION OF HETEROCHROMATIN (Casperson *et al*. 1968, 1969 *a* and *b*, Vosa 1970, Adkisson *et al*. 1971, Vosa and Marchi 1972).

Material: Squashes of plant somatic cells; animal somatic chromosomes in squashes or spread free cells; squashes of salivary glands and neural ganglia of Diptera. Quinacrine dihydrochloride is sold under the trade name of Atebrin in Great Britain; Quinacrine mustard is difficult to obtain. They are used, respectively, as 0·5–1% and 0·005% solutions either in water or absolute alcohol according to convenience.

A. Pre-treat tissues or free cells according to usual practice *viz*., plant tissues with colchicine, animal cells with colcemid and/or hypotonic solution.

B. Fix salivary glands and ganglia in 45% acetic acid for 3 min and other tissues or cells for usual periods, either in acetic-ethyl alcohol or acetic-methyl alcohol as 1 : 3 solutions.

C′. Prepare salivary glands and ganglia as squashes in 45% acetic acid; macerate and squash plant tissues in 45% acetic acid; spread free animal cells on slides by drying in air.

C″. Separate cover slips from slides of squash preparations by Quick-Freeze Method (Schedule 8) and store slides bearing the cells in absolute alcohol, or dry in air after removal of cover slip.

D. Stain for 5–20 min, the shorter time is sufficient for some animal cells.

E′. Rinse briefly 2–3 times in distilled water or absolute alcohol, according to whether aqueous or alcoholic solutions of the stain are used.

E″. Dry the preparation in air if alcohol is used for rinsing.

F. Mount in water and ring cover slip with rubber solution to prevent drying out. The preparations remain satisfactory for 2–3 weeks if stored in a cool place.

Note. According to Vosa and Marchi (1972) an enhancement of fluorescence was obtained in the bands of the chromosomes of *Vicia faba* when stained preparations were mounted in buffer solution (pH 7·8) instead of water.

SCH. 20. CARBOL FUCHSIN METHOD.

(For human chromosomes (Carr and Walker 1961) and general stain for chromosomes in autoradiography before application of stripping film (Bianchi *et al*. 1964).

Fixatives: Carnoy, acetic-alcohol or methyl alcohol.

Material: Smears or squashes.

A. Stain in carbol fuchsin* 2–5 min.

B. Absolute alcohol; 2 changes, differentiate under microscope.

* See Formulae of Reagents.

C. Xylol; 2 changes, mount in 'De Pex'.

Note. For autoradiography dry in air after step B.

SCH. 21. TWEEN METHOD FOR NUCLEOLAR STRUCTURE.

(For revealing the organiser (loops) in the nucleolus, La Cour 1966).

Material: unfixed isolated nuclei of young endosperm and briefly fixed isolated nuclei from root tips.

(1) Endosperm.

A. Prick young developing seed with fine needle and expel endosperm on clean slide by slight pressure with the side of needle.

B. Apply a small drop of 0·1% aqu. soln. of 'Tween 80' (Polyoxethylene sorbitan monoleate) to the suspension of cells.

C. Leave 5–10 seconds before applying the cover slip.

D. Examine with phase contrast.

(2) Root tips.

A. Fix root tips 2–3 min in 2% neutral formalin.

B. Remove excess fixative by placing on filter paper.

C. Rinse in 0·1% 'Tween 80' soln.

D. Crush tips (1–2 mm) between two clean slides in a fresh drop of the 'Tween' soln., avoid shearing.

E. Apply 1–2 drops of the detergent soln. to the suspension of clean isolated nuclei.

F. Examine with phase-contrast.

Notes:

(1) The internal structure of the nucleolus will be seen only if the nucleus expands in the 'Tween 80' and this will not occur if the nucleus is in an intact cell. If the endosperm is fairly liquid or only beginning to become cellular, the detergent itself will strip off the cytoplasm.

(2) More detergent can be run under the cover slip to aid expansion and disperse the less permanent components of the nucleolus which envelop the organiser.

(3) Fixative can be run under the cover slip of treated preparations if fixation is desired for further studies.

(4) The fine threads of interphase chromosomes in the expanded nuclei become visible in aceto-carmine.

SCH. 22. AUTORADIOGRAPHY: STRIPPING-FILM TECHNIQUE (Pelc 1947, Doniach and Pelc 1950, Pelc 1956).

Kodak stripping plates (AR 10) can be handled in the dark room with a safe-light (Wratten—series No. 1) but avoid direct illumination as far as possible.

(1) PREPARATION OF AUTORADIOGRAPHS

A. Cut with a sharp scalpel or razor-blade through the film along the four sides of the plate, at about $\frac{1}{2}$ in from edge; then cut through the film to obtain strips about $1\frac{1}{2}$ in wide and about 2 in long.

B. Lift one of the strips at one corner with a scalpel and strip the film slowly from the glass. Avoid fast stripping since this may cause fogging. The stripped film consists of two layers, an upper layer of nuclear track emulsion and a supporting layer of pure gelatin which was next to the glass. It is important to remember which side is which.

C. Turn the strip over and float on dilute transfer solution kept at 18–21° C (Appendix II, *g*) in a clean glass dish, so that the strip lies flat on the surface emulsion side downwards. The film should be left to soak for about $2\frac{1}{2}$–3 min during which time it will expand considerably.

D. Take a slide from distilled water, specimen upwards, and slip into the dish underneath the floating film. Lift the film and slide together out of the water; carefully, so as to avoid air pockets forming between the specimen and film. Drain and dry in front of a fan at room temperature, about 5–10 min).

E. Store in slide boxes in absolute darkness, preferably at a temperature of 4° C.

F. Develop for 5 min in D. 19b developer (see Formulae of Reagents) at 18° C. Fix in 0·25 M strength acid fixer (at 18° C) 9–10 min.

G. Wash in tap-water for 30–60 min with a final rinse in distilled water. If the tap-water is not clean, wash by repeated changes of distilled water at 18° C.

H. Immediate drying of the slides may be desired, *viz*. where tissues have been previously stained and require mounting in 'Euparal', or where slides are to be stored for staining at a later date. They are best dried in front of a fan in an atmosphere free from dust.

Notes:

(1) When handling strips of film the fingers should be dry. The strips should be held at the extreme corners since it is important not to damage the surface of the emulsion. Stripping may be difficult in warm and humid conditions but the difficulty can be overcome if plates are dried in a desiccator.

(2) An 'ovenware' type of glass dish 10 in long 6 in wide $1\frac{3}{4}$ in deep is very suitable for expanding the film. With careful timing four strips can be expanded at a time. The distilled water used for this purpose should be changed once the full light has been turned on. On no account use a dish that has been previously used for developing or fixing. Coplin jars are suitable for handling slides for developing and fixing. Some types of wooden slide boxes are unsuitable for storing autoradiographs before processing; they give off fumes causing fogging.

(3) Slipping or blistering of the film can either be due to unsuitable coating of the slides with egg albumen or to insufficient difference between the temperature of water used for expanding the film and the solutions used afterwards, or to high concentration of acid photographic fixer, or stains.

(2) STAINING OF AUTORADIOGRAPHS

Note. The temperature of all solutions should be below 18° C. Slides which have been stored dry require soaking in water (below 18° C) for 30–60 min before staining. All stains should be filtered to prevent precipitates from adhering to the wet film. (See Pelc 1956 for other special methods of staining.)

Method 1. Ehrlich's haematoxylin and eosin, for general morphology and nuclei after removal of DNA with D N-ase.

A. Stain in Ehrlich's haemalum 30 min.
B. Differentiate in 0·2% aqueous HCl 30–60 seconds.
C. Blue in running tap-water 30–60 min.
D. Stain in 1% eosin 3–5 min.
E. Wash till film has lost most of eosin.
F. Dry in air and mount in 'Euparal'.

Method 2. Toluidine blue, for chromosomes.
 A. Stain in 0·5% aqueous toluidine blue 30–60 seconds.
 B. Wash in water 5 min.
 C. Dry in air and mount in 'Euparal'.

Method 3. Methyl-green-pyronin for nuclei, nucleoli, cytoplasm and for RNA test where tissues have been previously treated in R N-ase.
 A. Stain in methyl-green-pyronin 30 min.
 B. Rinse in distilled water.
 C. Dry quickly in air and mount in 'Euparal'.

Method 4. Metanil yellow for nucleoli where nuclei have been stained in Feulgen before autoradiography.
 A. Stain in 0·03% aqueous metanil yellow 30 min.
 B. Wash in running tap-water 2 min.
 C. Rinse in distilled water.
 D. Dry film quickly in air and mount in 'Euparal'.

SCH. 23. FLOATING CELLOPHANE METHOD FOR POLLEN (La Cour and Fabergé 1943, Narasimhan 1963).

Uses: Pollen germination tests, or pollen tube mitosis.
(For sugar percentages see Ch. 10.)
 A. Wet cellophane* square (about 2 × 2 cm) in sugar solution. Remove excess by blotting.
 B. Float square on a large drop of sugar solution in a petri dish. Keep upper surface free of solution.
 C. Sow ripe pollen, by dusting on the square. Replace petri dish lid. Keep in a temperature of approximately 20° C.
 D. Examine periodically under microscope.
For germination studies, staining is not generally necessary. The following methods are suitable for study of pollen-tube mitosis:
 (i) Fix and stain in acetic-orcein 5 min by covering the square face upwards with the solution on a slide.
 (ii) Fix in acetic alcohol 2–24 hours. Stain in leuco-basic fuchsin after 6 min hydrolysis at 60° C.
N.B. Permanent preparations can be made with both methods.
  Rinse in 50% alcohol 2 seconds.
  Proceed as from J in Schedule 5 or from A in Schedule 8.

SCH. 24. BLOOD CULTURE METHOD FOR HUMAN CHROMOSOMES IN LEUCOCYTES FROM WHOLE BLOOD (Hungerford 1965).

Culture medium:
 Earle's balanced salt solution, containing at double strength Earle's basal amino acids and vitamins, is adjusted to pH 7 with 7·5% $NaHCO_3$ and supplemented as follows: glutamine 2 mM; penicillin, 100 units/ml; streptomycin, 0·1

* The cellophane must be not thicker than 0·04 mm and of the non-waterproof type. To accumulate metaphases 0·05% colchicine can be combined with the sugar solution (Swanson 1942) or acenaphthene crystals can be scattered in the petri dish.

mg/ml and phenol red, 0·007 mg/ml. Foetal or new-born agammaglobulin bovine serum and phytohemagglutinin M (Difco) are then added to make 15% and 2% respectively of the final volume, and finally 20 000 U.S.P. units of heparin sodium to each litre of complete medium.

Large quantities of the medium can be stored as 5 ml lots in 60 × 25 mm screw-cap vials at −25 to −30° C and be ready for immediate use when thawed. No deleterious effects have been detected after several weeks storage.

Culture procedure:

A. Each vial of media is inoculated with 0·2 ml of whole blood obtained by venipuncture. Successful cultures have also been obtained with 0·1 ml inocula collected in capillary pipettes, after finger puncture.

B. Cultures are incubated in closed vials in a gas phase initially of room air for three days at 37° C. Colcemid is added to give a concentration of 0·002 mg/ml for the final 3–5 hr.

Cytological preparation:

C. Transfer each culture by pipette into a graduated 15 ml conical-tip centrifuge tube and sediment the cells by centrifugation at 100 × g for 4 min. Siliconed pipettes and other glassware should be used to guard against loss of cells by adhesion to glass surfaces.

D. Remove supernatant and re-suspend the cells in 0·075 M KCl (warmed to 37° C) containing 16 U.S.P. units of heparin sodium and incubate for 10 min, including the time required for subsequent centrifugation.

E. After centrifugation replace the hypotonic KCl with 3–5 ml of freshly made fixative (3 parts methyl alcohol : 1 part glacial acetic acid).

F. Immediately disperse the pellet of cells into the fixative by gentle agitation with a pipette.

G. Centrifuge and re-suspend the cells twice more in freshly made fixative, allowing an interval of at least 15 min between changes, during which time the suspended cells are kept at icebox temperatures. The interval may be extended to several hours or overnight, providing freshly prepared fixative is used for the final change.

H. Let 3 evenly-spaced drops of suspension fall on a clean slide. Allow the fluid to spread to its maximum extent, then, as it starts to contract, blow gently until dry.

I. The preparation can be stored dry for later examination, or for later use in various banding techniques, or stained immediately either with lactic-acetic orcein or toluidine blue.

Notes:

(1) Mitoses are consistently more frequent in cultures from some donors when the incubation temperature is elevated to 38–39° C and from some at temperatures below 37° C, in rare instances.

(2) Step (H) deviates from the original, but the method of preparing air-dried spreads is rather a matter of choice.

(3) Cold treatment of the culture at 4° C can also be employed to accumulate metaphases (Heuser and Razari 1970).

SCH. 25. AIR-DRIED CHROMOSOME PREPARATIONS FROM BONE MARROW (Tjio and Whang 1962) (Suitable for Mammals, Amphibia, and Birds).

Solutions required for pre-treatment:

(a) 6·6 mM sodium phosphate buffer (pH 7) containing 0·85% (w/v) sodium chloride to which is added either colchicine or colcemid 0·001 mg/ml.

(b) Sodium citrate 1% in distilled water.

A. Aspirate about 0·5 ml sternal iliac crest or tibial marrow and drop immediately into 2–3 ml of solution (a). If the pieces of marrow are large, free them as much as possible of blood clots and transfer to fresh solution. Leave for 1–2 hours at 20–30° C.

B. Centrifuge at room temperature 4–5 min at 400 rev min^{-1}. Remove supernatant and add 2–3 ml of solution (b). Shake to loosen cells and leave 30 min at room temperature.

C. Centrifuge 4–5 min at 400 rev min^{-1} and remove supernatant. Add 5 ml freshly prepared acetic alcohol (1 : 3), re-suspend by agitation and leave for 2–5 min. Repeat the procedure a second time.

D. Centrifuge again and remove supernatant, add a drop of fresh fixative and agitate to re-suspend the cells.

E. With a Wintrobe or similar pipette put 1 or 2 mm droplets of the cell suspension on chemically clean slides. Blow on each droplet gently as soon as it is in place to assist in spreading and drying. Leave to dry thoroughly.

F. The preparations can be stained either in acetic-orcein, Feulgen or Giemsa, or stained and mounted in the combined toluidine blue stain–mounting medium (Appendix II).

G. After acetic-orcein, the preparations can be made permanent by the quick-freeze method Schedule 8; after Feulgen as in Schedule 6; after Giemsa the preparations can be air-dried and mounted in 'De Pex' or other neutral mounting media.

SCH. 26. MEIOTIC PREPARATIONS FROM MAMMALIAN TESTES (Ford and Evans 1969).

(For stages of meiosis from late diplotene onwards of mice and other rodents, but adaptable to larger mammals including man.)

A. Kill rodent, dissect out testes and place them in a 2 in petri dish containing isotonic (2·2%) trisodium citrate solution.

B. Pierce tunica to expose tubules, then swirl in isotonic solution to wash away fat, allowing 15 seconds for the operation.

C. Transfer to another dish containing about 3 ml of fresh isotonic solution and tease tubules with curved forceps. Transfer contents of dish to a 4 ml centrifuge tube and pipette gently. Allow 10 min for whole step.

D. Spin 5 seconds in centrifuge to sediment larger tubule segments. Transfer supernatant to clean 4 ml centrifuge tube and discard sediment.

E. Spin 5 min at 500 rev min^{-1}, discard supernatant and re-suspend pellet in minimum volume of residual supernatant.

F. Slowly add 3 ml of hypotonic (1%) trisodium nitrate solution to tube, meanwhile flicking it with the forefinger, then divide between two Dreyer tubes (narrow conical tipped tubes, about 2 ml capacity).

G. Spin 5 min at 500 rev min^{-1}. Pour off supernatant, let tubes stand 1 min and then remove with a fine pipette the residual fluid that drains down the walls. Make up fixative (absolute alcohol 4·5 ml; acetic acid 1·5 ml; chloroform 0·1 ml) while the material is in the centrifuge.

H. Flick tube so as to disperse pellet as a dense suspension. Allow two drops of fixative to fall directly on to cells and then flick tube vigorously. Add more fixative until about threequarters full, meanwhile continuing to flick the tube. Spin 3 min at 500 rev min^{-1}. Change fixative, then spin and change fixative twice more. Allow 15 min for whole step.

I. The final suspension of cells in fixative should be dilute. Let it stand for 1–3 hours before making preparations. After standing, it may be necessary to re-suspend cells before proceeding.

J. Take up cell suspension into a fine bore pipette (made from 4 mm diameter soft glass tubing; teats can be made from short lengths of polythene tubing, the ends being sealed by heating followed by pressure with pliers).

K. Let 3 evenly-spaced drops fall on a clean slide. Allow the fluid to spread to its maximum extent and start to contract (*i.e.* when by the light of a bench lamp interference colours are seen to appear in the remaining thin film of liquid), then blow gently until dry. Repeat until there are a sufficient number of cells on the slide, as judged by phase contrast examination.

L. The preparations can be stored dry for later examination, or stained immediately either with lactic-acetic-orcein or toluidine blue. The former is preferred for its sharpness.

Notes:

(1) The number and quality of the cells in the final suspension will be dependent on the extent and vigour of teasing and subsequent pipetting (steps B and C). Too much will give a suspension containing numerous small tubule fragments; too little will result in many spermatogonial cells being lost within the discarded tubules.

(2) The optimum concentration of the hypotonic solution and the length of exposure to it will probably differ with the species and will have to be determined empirically. Freshly prepared citrate solutions should be used and it is possible that other hypotonic fluids (*e.g.* potassium chloride, Hungerford 1965) might give improved results with some species. Better preparations of testicular material of man have been obtained by teasing the tubules in Hank's solution and then transferring to 1·5% trisodium citrate.

(3) When transferring the fixed cells to the slides, the drops of fixative containing them should spread evenly on the slide, forming a film with an even uncrinkled edge. Do not use siliconed slides and beware of some detergents used in cleaning them, as these may also adversely affect spreading. A single gentle exhalation should suffice to dry the film.

(4) It is sound practice to make one preparation immediately after the final change of fixative, in order to assess the cell concentration and, if necessary, adjust it.

SCH. 27. IDENTIFICATION OF SISTER CHROMATIDS BY DIFFERENTIAL STAINING

(Ikushima and Wolff 1974, Perry and Wolff 1974, as applied to Chinese hamster ovary cells).

Culture medium and procedure:

Cells are grown in McCoy's (5a) medium supplemented with 15% foetal calf serum, in prescription bottles at 37° C in 5% CO_2 in air.

Treatment and procedure:

A. Under dim light, sow cells in 5 ml of medium to which is added 0·003 mg/ml BUdR or IUdR (final concentration 10 μM), in 2-oz. bottles.

B. Incubate for 24 hr in dark at 37° C.

C. Add colcemid (at a final concentration of 2×10^{-7}M) and leave a further 2 hr in dark.

D. Collect cells by shaking and treat for 8 min with 0·075 M KCl (pre-warmed at 37° C) in dark.

E. Harvest cells by centrifugation (100 $\times$ g for 4 min) and fix for 30 min in methanol–acetic acid (3 : 1).

F. Place small drops of cells in fixative on clean microscope slides and allow to dry in air.

G. Stain fixed preparations for 12 min in Hoechst (33258) stain at a concentration of 0·5 μg/ml.

H. Rinse briefly with deionised water, mount and ring the cover slip with 'Holdtite' rubber cement to prevent evaporation.

I. The sister chromatids should now fluoresce differentially (*cf.* Latt 1973).

J. Allow stained preparation to age for 24 hr, remove cover slip and incubate specimen for 2 hr at 60° C in either 2 $\times$ SSC (0·3 M sodium chloride–0·03 M trisodium citrate) or in distilled water.

K. Stain for 30 min in 5% Giemsa solution (Gurr's R66) in phosphate buffer pH 6·8.

L. Rinse briefly in water, dry in air and mount in 'De Pex', or other suitable mountant.

Notes on staining:

The fluorescence given by the Hoechst stain fades rapidly. A similar more stable fluorescence can, however, be obtained if in Step G the cells are stained for 5 min in 0·005% acridine orange in Sörenson's buffer pH 6·7. The differential fluorescence is then not readily distinguished until the cells are exposed to exciting light passed through a BG 12 filter, and it is only after such exposure that subsequent differential staining of sister chromatids with Giemsa is obtained. If, however, fading of fluorescence occurs because of over-exposure to exciting light, the subsequent staining with Giemsa is then very weak. Similarly, if cells stained with the Hoechst stain are exposed to exciting light, they do not stain as well with Giemsa as unexposed cells.

General notes:

(1) Treatment with SSC saline (Step J) may lead to G-bands being seen.

(2) The technique can be applied to other mammalian cells in culture and no doubt to testes of Orthoptera etc. after injection of BUdR and probably with suitable adaptation to plant roots and PMC. A knowledge of the duration of the mitotic cycle is essential.

APPENDIX IV

Catalogue of Implements

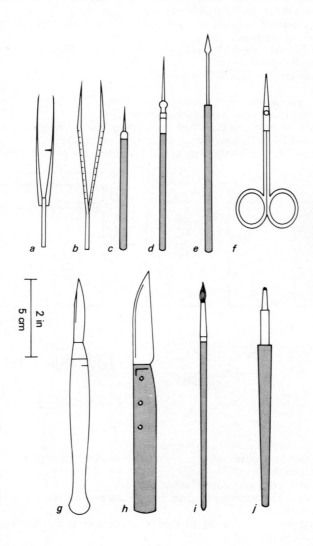

FIGURE 2

(i) *Metal Instruments* (Fig. 2).

(*a*) Fine forceps for cutting off root tips and dissection.

(*b*) Coarse forceps for holding slides.

(*c*) Bone needle holder for tapping out squashes (blunt end) and dissection (pointed end).

(*d*) Aluminium mounted needle with lancet blade and soldered safety-razor blade attachment for cutting tissues or raising cover slips.

(*e*) Plain mounted needles, one iron for teasing aceto-carmine, one nickel-plated for other purposes.

(*f*) Spear-shaped mounted needle with two cutting edges for dissection and cutting of small tissues.

(*g*) Fine scissors for animal dissection.

(*h*) Scalpel.

(*j*) Flat-honed potato-knife scalpel for smearing.

(*k*) Mounted rimming-rod for sealing cover slips.

(*l*) Camel-hair brush, one needed for holding microtome ribbons and transferring material from fixative, etc., another for cleaning eyepieces.

(*m*) Diamond (or carborundum) pencil for engraving numbers on slides.

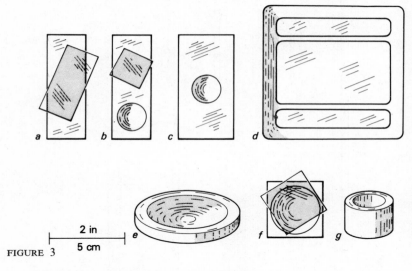

2 in

5 cm

FIGURE 3

(ii) *Glass and Earthenware Implements* (Fig. 3).

(*a*) Slide and long cover slip for ribbons or smears. All slips should be grade 0 (0·075–0·1 mm thick).

(*b*) Slide and cover slips for squashes, etc.

(*c*) Well-slide for warming fixative and tissues for maceration, or for fixation of very small objects.

(*d*) Ridged smearing dish 3½ in square, earthenware, for fixation, bleaching and, if necessary, staining.

(*e*) Solid, flat-bottomed watch-glass for embedding or for dissection of small animals in Ringer or wax.

(*f*) Solid watch-glass with cover for fixation, etc.

(*g*) A 'van Tieghem' cell for hanging drop preparations (Ch. 3 and 10).

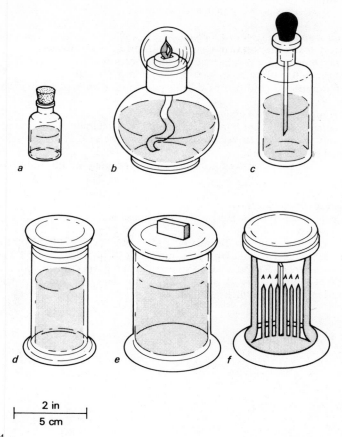

2 in
5 cm

FIGURE 4

(iii) *Glass Bottles and Lamp* (Fig. 4).
   (*a*) Fixing bottle, stoppered.
   (*b*) Spirit lamp for heating acetic squashes and smears.
   (*c*) Pipette bottle for acetic stains, etc.
   (*d*), (*e*) Jars for taking up slides with cover slips in making squashes and smears permanent. Cover slip holders may be made of aluminium wire.
   (*f*) Ridged jars for staining.

(iv) *Other General Items*.
   (*a*) Dark bottles, ground stoppered for H_2O_2 and OsO_4.
   (*b*) Pipettes, graduated and ungraduated.
   (*c*) Measuring cylinders, 10–100 ml.
   (*d*) Water vacuum pump.
   (*e*) Small hand pump for field work.
   (*f*) Dissecting pins, various.

(*g*) Dissection board.

(*h*) Small petri dishes for pollen germination.

(v) *Special Apparatus for Paraffin Sections.*

Sections demand the following extra apparatus:

(*a*) *A thermostatic oven* is needed for three purposes:

   (i) Embedding material in paraffin wax, temp. 50°–60° C.

   (ii) Evaporating chloroform during penetration of material by wax, temp. 50°–60° C.

   (iii) Hydrolysis for Feulgen reaction, temp. 60° C.

(*b*) *A thermostatic hot plate* is needed for stretching and drying paraffin ribbons and slides with cover slips mounted on balsam. The top of the oven can be used as a substitute.

(*c*) *A microtome* is needed to cut sections ranging in thickness from 3 to 40 μm.

(vi) *Drawing Implements.*

(*a*) H pencil, for drawing the outline by camera lucida.

(*b*) Indian ink, fixed.

(*c*) Pen nib, fine pointed; Brandauer 515 is the best design, 24 mm long and 4 mm broad, with a triangular tip 10 mm long. With such a nib a line can be widened to a streak 5 mm broad, the point making an accurate edge.

(*d*) Bristol board; 3-ply is thick enough for handling and thin enough for cutting and mounting on thick squared paper.

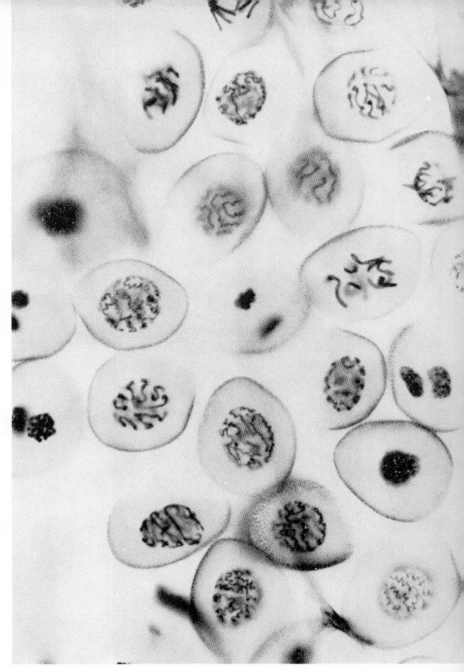

First PG mitosis in *Paris quadrifolia*, Liliaceae, showing all stages from prophase to telophase. $n = 10$ (*cf*. Darlington 1937, 1941)
2BE—CV smear, 8 mm objective, × 800.

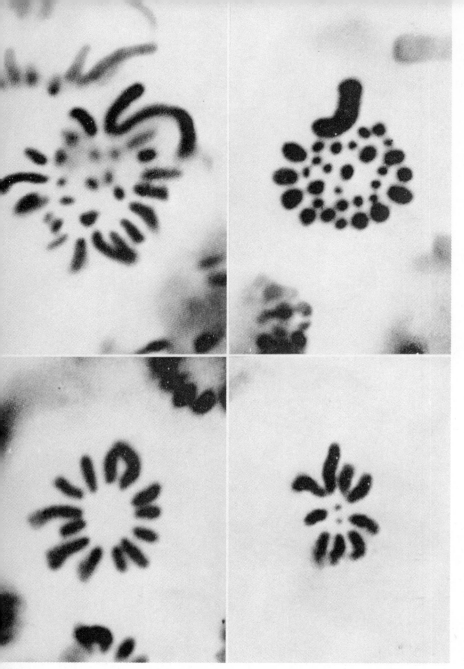

Polar views of metaphases in the spermatogonial cells of the testes of insects.

1 and 2, mitoses in young and old follicles in *Leptophyes punctatissima*, Orthoptera; 30 one-armed autosomes and one two-armed X (no Y).

3, *Tetrix bipunctata*, Orthoptera; 12 autosomes and one indistinguishable X (no Y).

4, *Drosophila miranda*; the chromosomes in order of size are: X_1, Y, X_2 and three pairs of autosomes.

1, 2 and 3, sections cut at 20 μm, fixed 2BD, stained CV.

4, acetic-lacmoid squash, prep. and photo by P. C. Koller.

1 and 2, $\times$ 3500, 3, $\times$ 4000, 4, $\times$ 10 000.

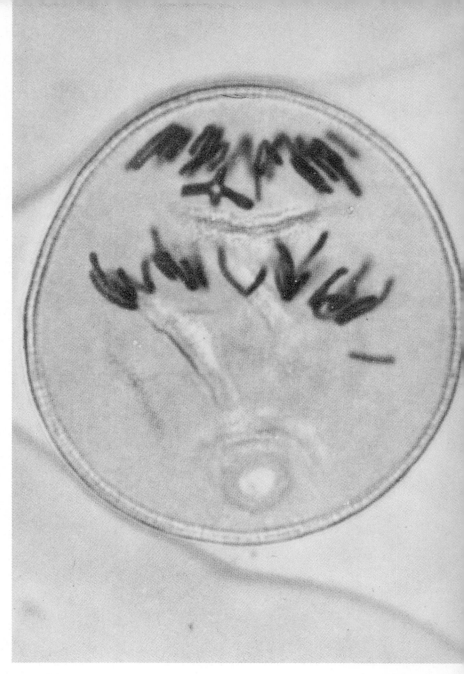

The birth of a B chromosome: anaphase of the first pollen grain mitosis in *Agropyron desertorum* (normally $n = 14$). One extra telocentric chromosome stands to the right of each group. One potential B chromosome has its division obstructed. Consequently both chromatids will be included in the group which is stopped by the pollen grain wall and is destined to form the generative nucleus. Note the pore at 6 o'clock.

Technique: Carnoy (6 : 3 : 1) 45 min, HCl–95% alcohol (1 : 1) 15 min, Carnoy 10 min, Hydrolysis 10 min, Feulgen 1 hr. Distilled water 30 min, Squashed in 45% acetic acid. Photographed fresh. Preparation by J. B. Hair.

× 2000.

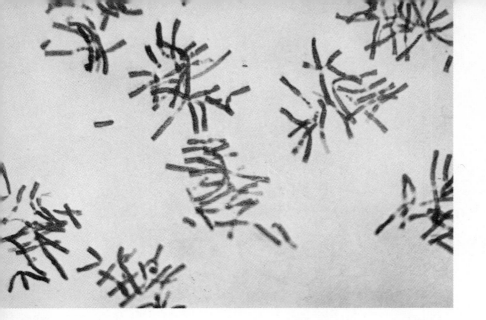

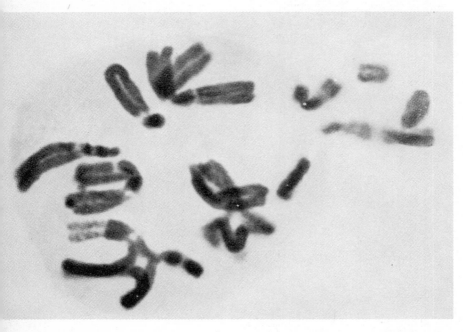

PLATE IV

Metaphases from squashes of *Trillium grandiflorum* ($2n = 10$) after cold treatment during the preceding resting stage (Darlington and La Cour 1945). 1, Endosperm, 2, RT.

1, Synchronised mitoses ($3n = 15$) : heterochromatic segments differentiated after 4 days at $0°$ C. Acetic-alcohol fixation, Feulgen staining (*cf*. Rutishauser and La Cour 1956).

2, X-rayed with 135 R and kept for 48 hr at $24°$ C and 96 hr at $0°$ C. Heterochromatin differentiated, chromatid and chromosome breaks (B' and B'') in the euchromatin. B' have been followed by R' at 5 and 7 o'clock, B'' have been followed by sister reunion of the broken ends both in the centric fragment (11 o'clock) and in the acentric fragment (2 o'clock).

1 and 2, 2BD-Feulgen squashes, $\times$ 1700.

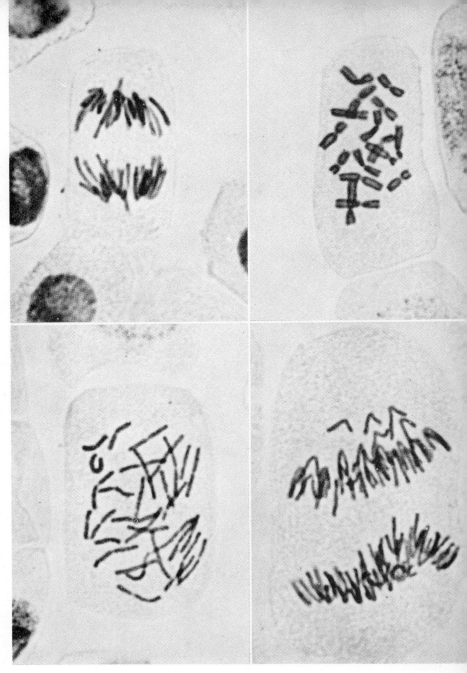

RT squashes of *Allium cepa* showing the action of a 0·05% aqueous solution of colchicine, on mitosis in the root (Thomas 1945).

1, Normal untreated diploid anaphase.

2, Metaphase 12 hr after treatment at 24° C showing the effect of spindle inhibition in delaying metaphase and over-contracting the chromosomes.

3, Abortive anaphase from the same preparation.

4, Normal anaphase after removal of the drug in a tetraploid cell produced by failure of mitosis (one day after 3).

All preparations by Prof. P. T. Thomas, 2BD-Feulgen, × 1000.

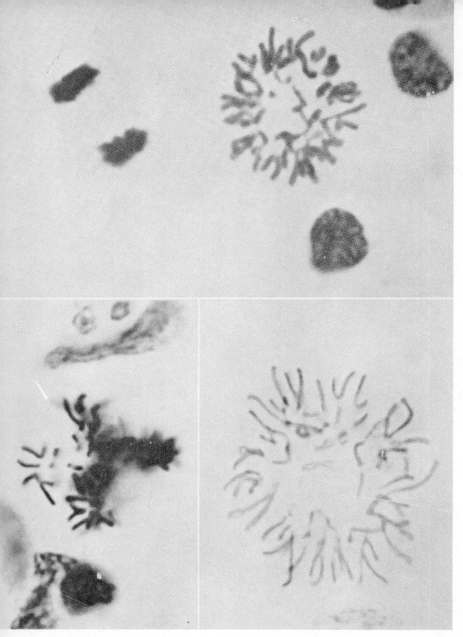

Mitosis in bone marrow showing the red and white blood precursor cells in man, $2n = 46$ (La Cour 1944).

1, Left, anaphase in a pre-erythroblast with heavy charging of DNA. Right, metaphase in a pre-myelocyte showing a lighter charging of DNA on chromosomes.

2 and 3, Corresponding metaphases from a man with pernicious anaemia, same preparation, showing excessive heavy and light charging due to exaggerated differentiation of the two types of cell. The heavy charge goes with supercontraction and a multi-polar spindle. The light charge with a hollow spindle and an abortive anaphase.

1, Leishmann, re-stained in Feulgen preparation by Dr H. Levy, $\times$ 2500.

2 and 3, Film air-dried 30 seconds (followed by blood fixative, Ch. 8) and Feulgen staining, $\times$ 3000.

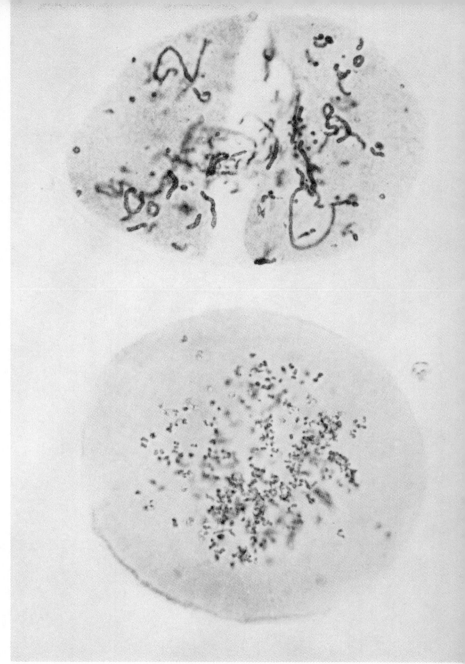

Spontaneous chromosome breakage.

1, Complete PMC at second anaphase in *Tulipa orphanidea* following fragmentation at first prophase. Numerous acentric chromosomes from breakage and ring chromosomes from reunion (Darlington and Upcott 1941*b*).

2, Cell from a burst stomach tumour in man. Most of the chromosomes are centric fragments in a highly polyploid cell; the acentrics have been lost (Prep. and photo by Prof. P. C. Koller).

1, Aceto-carmine smear, × 1000.

2, Film air-dried, 5 min acetic-lacmoid, × 1800.

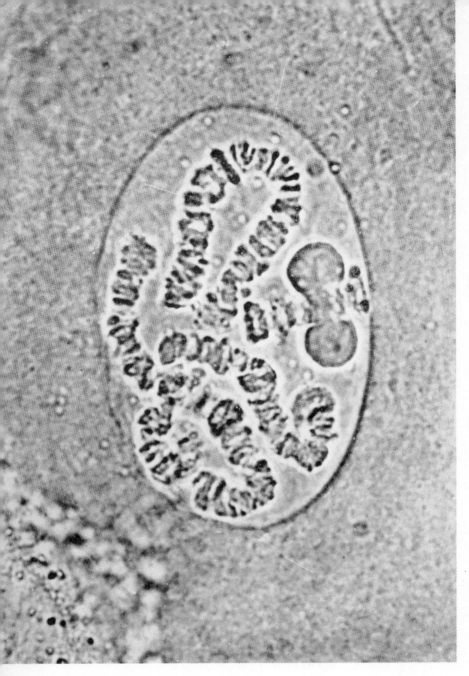

PLATE VIII

Living salivary gland nucleus of *Chironomus riparius*, mounted in liquid paraffin.
The nucleolus is attached to the fourth and smallest pair of chromosomes.
Prep. and photo by Dr A. M. Melland. 8 mm objective, × 600.

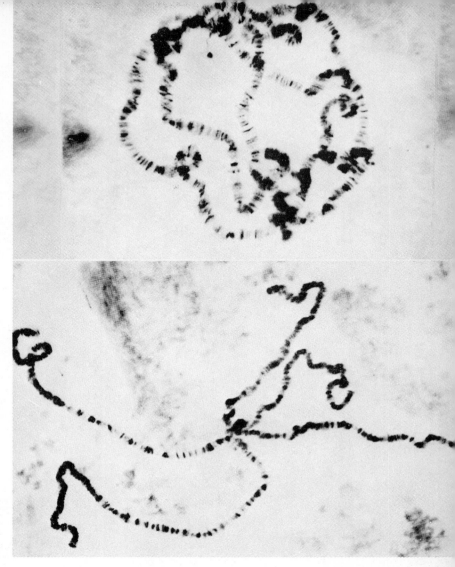

PLATE IX

Polytene chromosomes in the salivary glands of fly larvae. 1, *Drosophila* sp., nucleus unbroken; 2, *D. melanogaster*, nucleus broken.

All chromosomes, and in (1) the nucleolus as well, are attached to the fused body of heterochromatin.

1, Acetic-orcein (La Cour 1941), × 900. Process plate, no screen.

2, Aceto-carmine, × 500. Prep. and photo by Prof. H. G. Callan. 1 and 2, 8 mm objective.

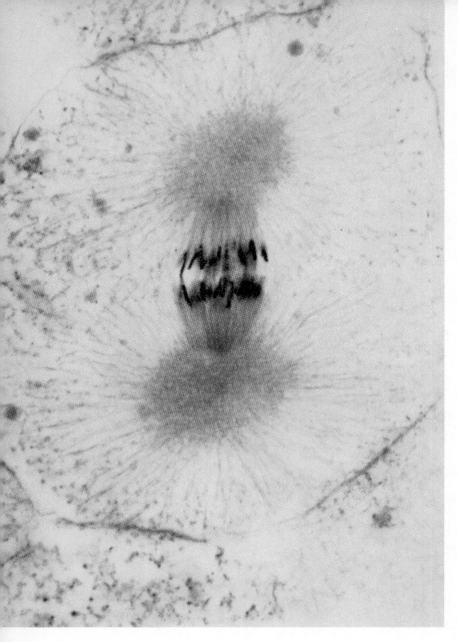

PLATE X

Cleavage mitosis in the morula of a teleostean fish, *Coregonus clupeoides*, in the middle of anaphase. Spindle structure revealed by slow fixation.

Section cut at 10 μm, $\times$ 4000. Strong Flemming, haematoxylin. Prep, and photo by Prof. P. C. Koller.

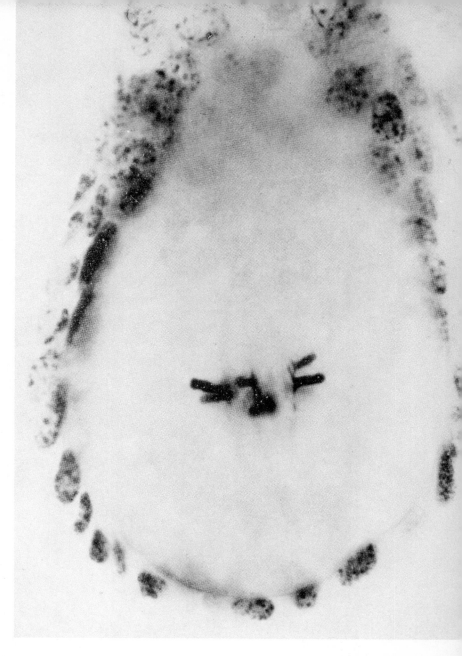

PLATE XI

First metaphase in EMC of *Fritillaria pallidiflora*, Liliaceae; the 12 bivalents have single chiasmata close to the centromere (*cf*. Darlington and La Cour 1941).

2BX—CV, section 40 μm, 8 mm objective, $\times$ 1200.

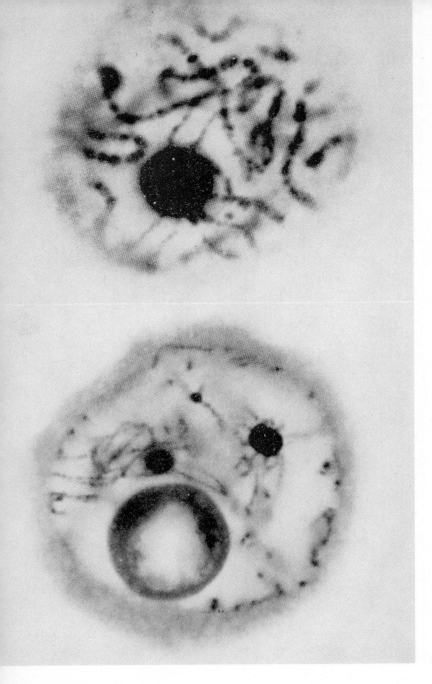

Prophase of meiosis (pachytene) of *Fritillaria* in species with and without heterochromatin.

1, *F. imperialis*, 2x, PMC, pairing of chromosomes attached to darkly stained nucleolus is not complete (Darlington 1935). Smear, Medium Flemming—CV, × 3000.

2, *F. pudica*, 3x, EMC, heterochromatin darkly stained, nucleolus darkened with osmic acid but unstained (Darlington and La Cour 1941a). Section at 40 μm, 2BX—Feulgen, × 3400.

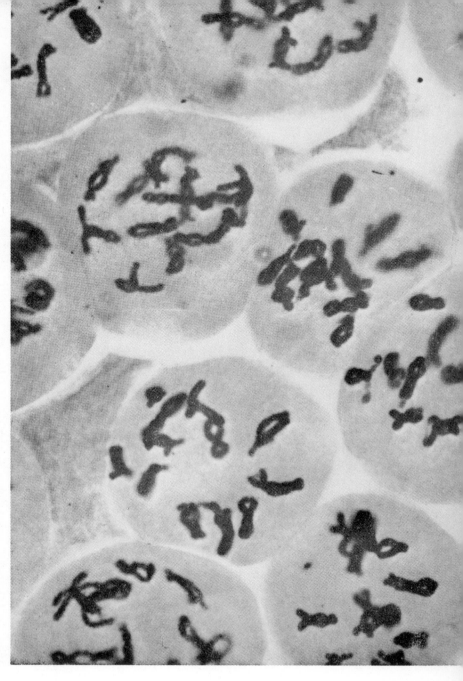

Pollen mother cells of *Fritillaria pallidiflora* ($n = 12$) at diakinesis to early first metaphase: note the varying numbers and positions of chiasmata and the beginning of co-orientation of the centromeres as the spindle develops (4 o'clock). Aceto-carmine. Unpub. × 1500.

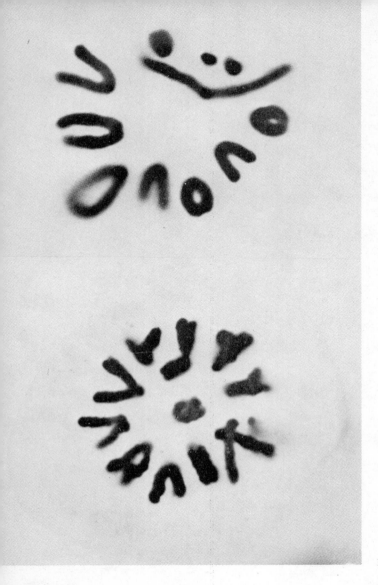

PLATE XIV

Polar views of first metaphase of meiosis, showing proximally localised chiasmata (*cf*. Plate XIII).

1, SMC of *Mecostethus grossus, n* = 11 (+ X, invisible).
2, PMC of *Fritillaria oranensis, n* = 12.
1, section at 30 μm, 2BD—CV, × 2700.
2, smear, 2BE—CV, × 1800.

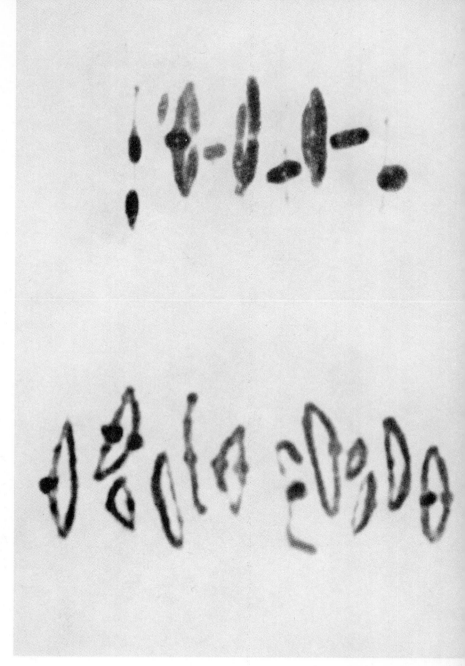

First metaphase in SMC of animals with terminally localised chiasmata (*cf.* Plate XVII).

1, *Chrysochraon dispar*, Orthoptera: $8^{II} + X$.

2, *Triton vulgaris*, Urodela: 12^{II}.

1, aceto-carmine squash. Prep. and photo by H. Klingstedt, $\times 1500$.

2, Acetic-alcohol—Feulgen squash. Prep. and photo by Prof. H. G. Callan, $\times 2400$.

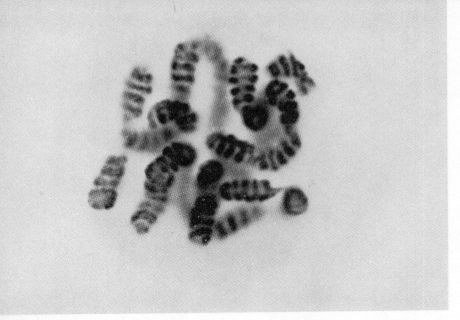

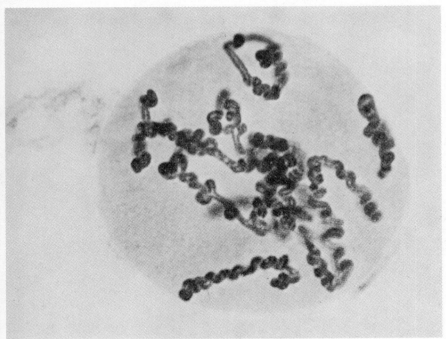

PLATE XVI

Spiral structure at first metaphase in *Tradescantia virginiana* PMC, $4x = 24$.

   1, Pre-treated with nitric acid vapour, medium Flemming—CV, $\times$ 2400.

   2, Acetic-lacmoid fixation following a shock due to sudden rise in temperature while growing, $\times$ 1800.

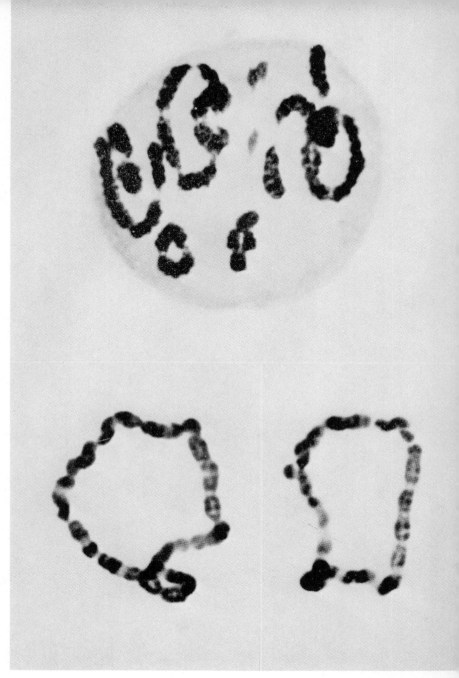

PLATE XVII

First metaphase in PMC smears of *Tradescantia* and *Rhoeo*, plants showing terminally localised chiasmata.

1, *T. virginiana*, $4x = 24 + 3$ fragments; 4^{IV}, 4^{II} (one with interstitial chiasma), $1 f^{III}$.

2 and 3, *Rhoeo discolor* with ring of 12, due to interchange hybridity: cold treated to show heterochromatin near the centromere.

1, Acetic—Bismarck Brown, $\times 1800$.

2 and 3, Acetic-lacmoid, 3 mm objective, $\times 2700$. Multigrade paper.

1, 2 and 3, Process plate without screen.

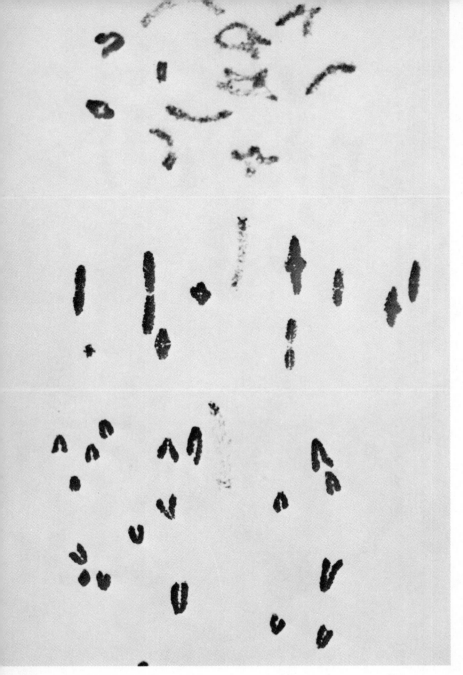

PLATE XVIII

Meiosis in the sperm mother cell of a locust *Pyrgomorpha kraussi* with 9^{II} + X + B. The unpaired X and B are both overstained at diplotene (top) and understained at first metaphase and also at anaphase when they are passing undivided to opposite poles. At diplotene there are 11 chiasmata, 5 terminal; at metaphase probably 10. The B chromosome in the anaphase has been cut off. Aceticalcohol, acetic-orcein (K. R. Lewis and B. John unpub.).

× 3000.

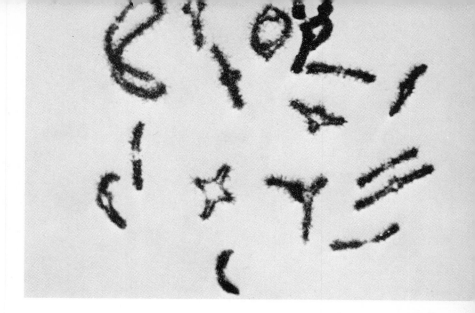

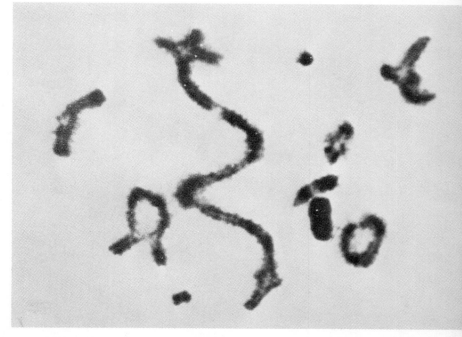

PLATE XIX

1, Diplotene in a tetraploid spermatocyte of *Pyrgomorpha kraussi* ($2n = 4x = 38$). Only one quadrivalent is present (a ring at 10 o'clock). The two hetero-pycnotic X univalents and two precocious bivalents are associated non-specifically (12 o'clock). Acetic-alcohol, acetic orcein (Lewis and John, *Chromosoma*, **10**, 589).

2, Diplotene in an artificial male hybrid from the cross *Eyprepocnemis plorans ornatipes* × *E.p. meridionalis*. These two races of African grasshoppers (or species as they deserve to be) have complements that are indistinguishable at mitosis but differ amongst other things by at least three interchanges. All the autosomes are acrocentric ($2n = 22 + X$). The cell shown has VIII + 6II + 2I + X. Acetic-alcohol, acetic-orcein (John and Lewis, *Chromosoma*, **16** : 308).

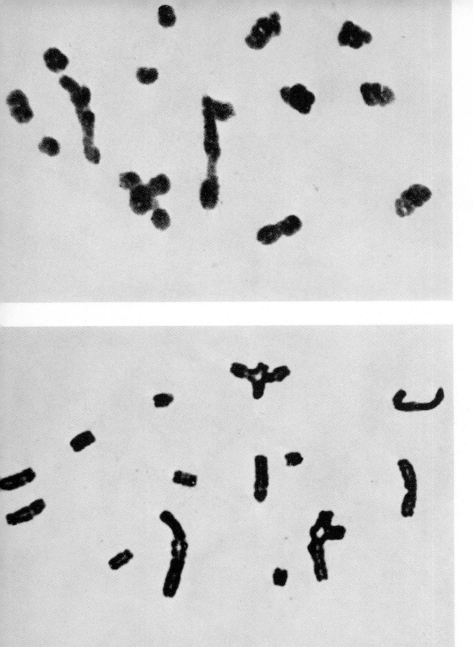

PLATE XX

Meiosis in *Narcissus* 'Geranium' (2n = 17) a first cross between *N. poeticus* (2n = 14) and *N. tazetta* (2n = 20).

The first metaphase (top) shows three unequal bivalents and eleven univalents. The lapse of chromatid attraction on which the onset of first anaphase depends, does not occur in this hybrid. A restitution nucleus is formed and the chromosomes reappear at second division with major coils undone but with the chiasmata still present. Three unequal bivalents, each with a single chiasma, together with 11 univalents are seen at second metaphase (bottom). Acetic-alcohol, aceto-carmine (K. R. Lewis and B. John, unpub.).

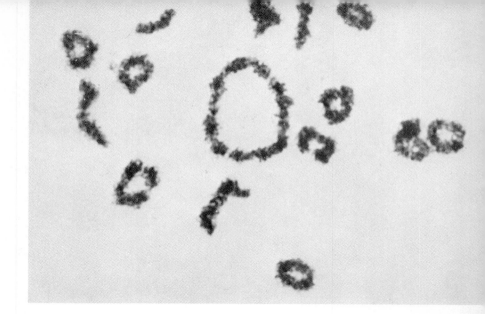

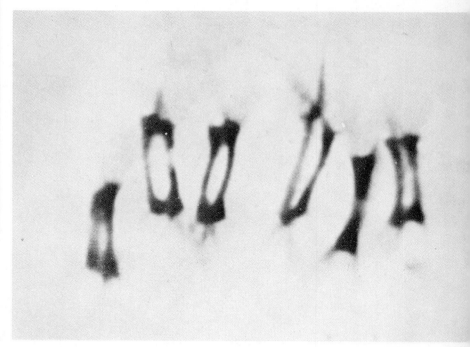

1, Diakinesis in the sperm mother cell of a cockroach, *Periplaneta americana* (2*n* = 32 and X unpaired at 5 o'clock). A ring of six shows heterozygosity for two interchanges. This stage is normally replaced by a 'stretch' due to centromere co-orientation which has here been suppressed by colchicine. Acetic-alcohol, acetic-orcein (B. John and K. R. Lewis 1958. *Heredity*, **12** : 185).

× 3000.

2, First metaphase in pollen mother cell of *Luzula maxima* (2*n* = 12). All six bivalents have two terminal chiasmata and non-localised centromeres. Acetic-alcohol, acetic-orcein (P. T. Thomas and K. R. Lewis, unpub.).

× 5600.

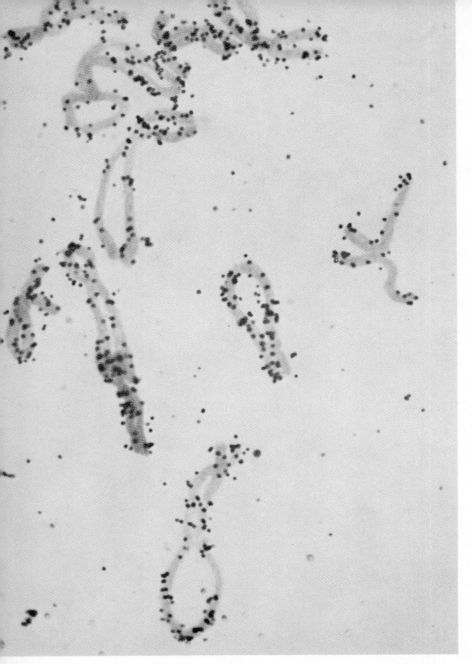

PLATE XXII

Autoradiation photograph: metaphase of mitosis in RT of *Scilla sibirica* (2n = 12) 24 hr after absorption of H³-thymidine, showing asynchrony of DNA synthesis in chromosomes. Fixation acetic-alcohol, toluidine blue staining after photographic processing. Photographed with a dark-blue filter to reduce contrast in chromosomes (La Cour and Pelc. unpub.).

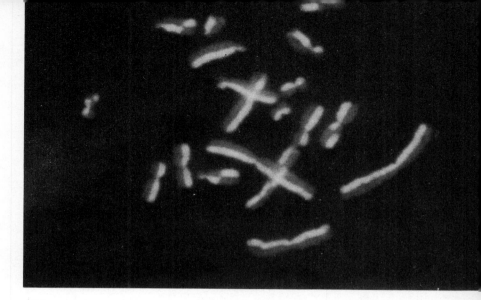

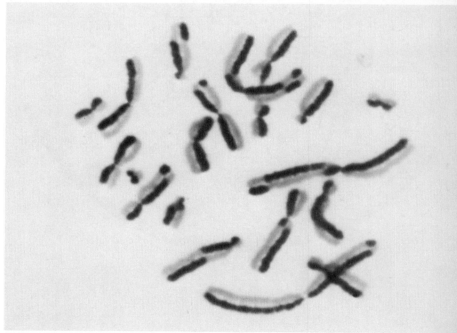

Metaphases in cultured fibroblasts of Chinese hamster (CHO line), following growth for two cell cycles in the presence of BUdR (10 μm), showing differentially stained sister chromatids and sister chromatid exchanges (see Schedule 27 and Perry and Wolff 1974. *Nature* **251**: 156).

1, Stained for 15 min with Hoescht 33258 and photographed with a Leitz fluorescence microscope, using UG1 and UG5 filters. The 19 chromosomes show 7 examples of sister chromatid exchange. The DNA of the chromatids with dull fluorescence is bifilarly substituted whereas only one strand of the DNA in the brighly fluorescing chromatids contains BUdR.

2, Stained with Hoescht 33258 and then exposed to light before staining with Giemsa. Regions that fluoresce brightly with Hoescht 33258 stain intensely with Giemsa. The 20 chromosomes show 12 examples of sister chromatid exchange. Prep. and photos by Dr P. Perry and Dr S. Wolff, × 1950.

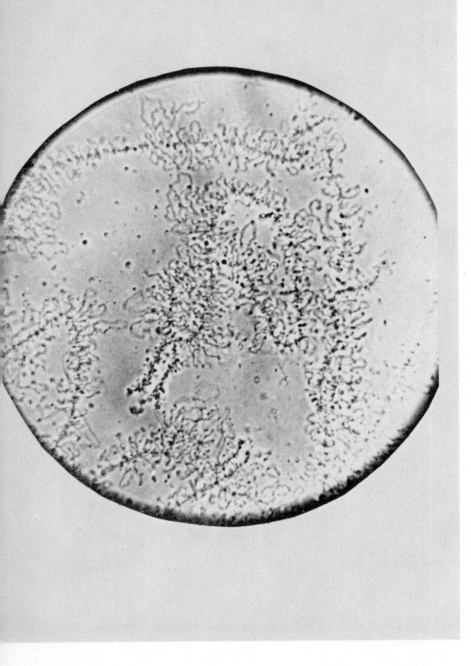

PLATE XXI

Lampbrush chromosomes: the shortest bivalent (XII) from diplotene in the living egg of *Triturus cristatus carnifex* showing loop filaments arising from the main axes of the chromosomes. Photographed over a × 40, 4 mm phase contrast objective without eyepiece magnification. Preparation by Prof. H. G. Callan (*cf.* Callan 1957. *Pub. Staz. Zool. Napoli.* **29**: 329), × 560.

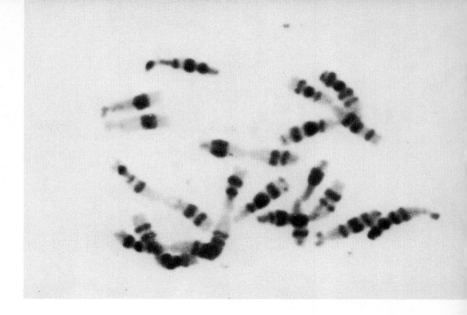

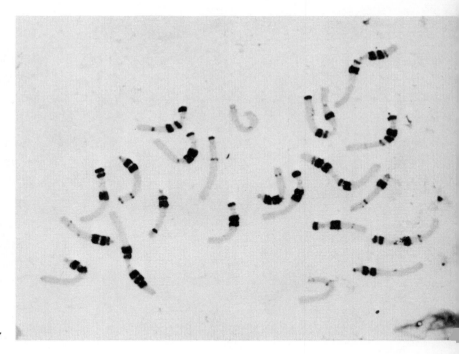

Illustrating G-banding in somatic metaphase chromosomes of two plants.

1, From a squash preparation of a root tip of *Anemone blanda* $2n = 16$, squashing facilitated by 20 min treatment in 45% acetic acid at 60° C prior to maceration (Schedule 18). Prep. and photo by G. E. Marks, $\times$ 2000.

2, From a squash preparation of a root tip of *Fritillaria lanceolata* $2n = 24$, squashing facilitated by treatment in 9 parts 45% acetic acid, 1 part N HCl (Schedule 18). Note that the intensely staining G-bands correspond to heterochromatic regions that in sections stain weakly with basic dyes (La Cour and Wells, 1974, *J. Cell Sci.* **14**: 505), $\times$ 800.

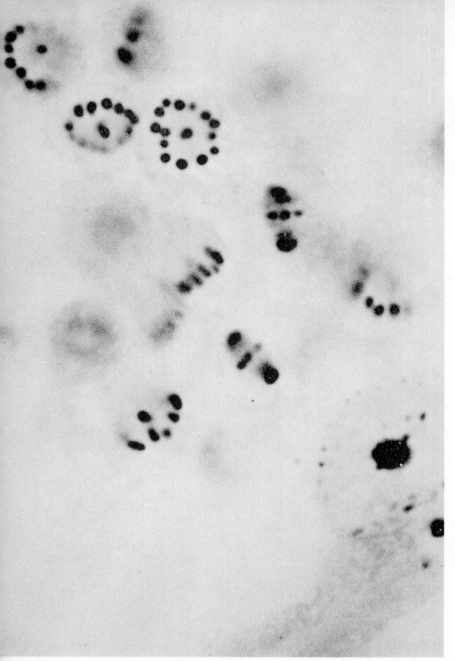

PLATE XXVI

Second metaphase in SMC of *Cimex rotundatus*, Heteroptera, $n = 14 + X_1X_2Y$. Polar and side views, show central co-orientation of the three sex chromosomes (*cf*. Darlington 1939).

Section at 16 μm, 2BD—CV, $\times$ 3000.

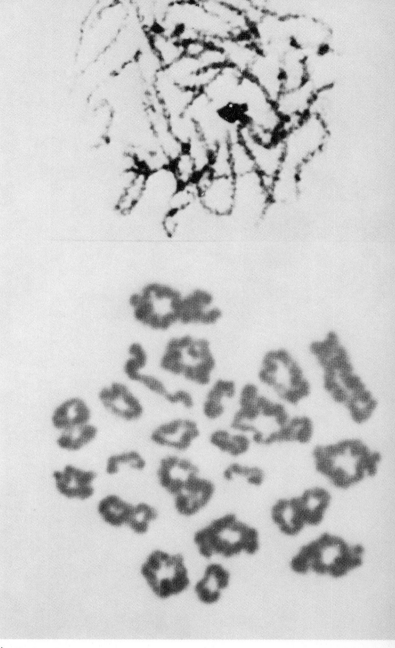

PLATE XXVII

Meiosis in man.

1, Pachytene in a human spermatocyte, showing heterochromatic sex bivalent, × 1100 *ca*.

2, First metaphase in the same individual. There are 21 bivalents including the sex chromosome pair, and an association of three elements, towards centre, which includes two non-homologous D-group chromosomes (probably 1, 3 and 14) and the product of their 'centric fusion', × 2700 *ca*.

Air-dried cells from a suspension fixed in acetic-alcohol after initial treatment in isotonic sodium citrate (Evans, E. P., Breckon, Gardford, C. E. 1964 *Cytogenetics* **3**: 289–294). Prep. by Dr J. R. Byrd, Medical College of Georgia, U.S.A.

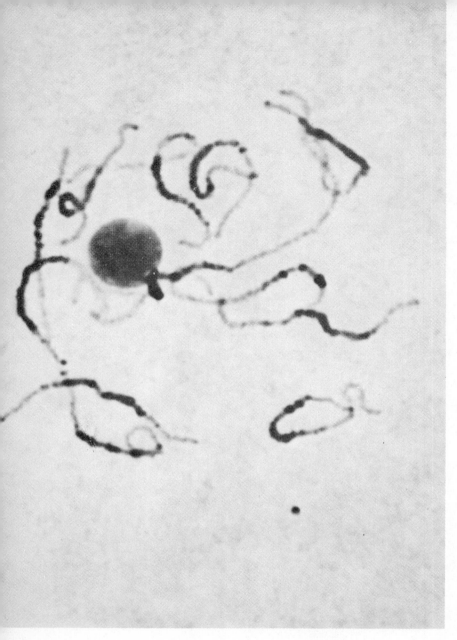

PLATE XXVIII

Tomato pachytene. Aceto-carmine after propionic-alcohol fixation and mordanting with iron alum, × 4000. Prep. by Brian Snoad.

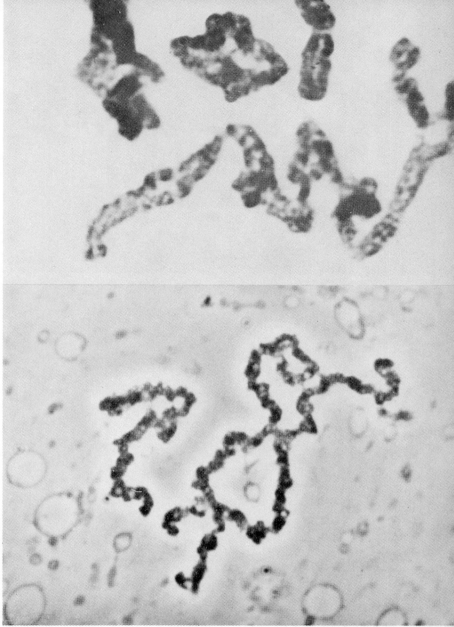

PLATE XXIX

1, First metaphase in the hybrid between *Tulbaghia violacae* and *T. dregeana*
2n = 12, showing the complete interchange chain of six chromosomes (2V + 2D
+ 6V + 4D + 4V + 6D) in which crossing over in the distal inversion of
chromosome pair 2V + 2D has taken place with consequent bridge and fragment.
There are two bivalents (pairs 2V + 2D and 5V + 5D) and the univalent
chromosome 3D. Chromosome 3V is out of the Plate. Acetic-orcein, × 3000 *ca.*
Prep. by C. G. Vosa (*Heredity* **21** : 675–687, 1966).

2, The three nucleolar organisers from a dispersed nucleolus of an unfixed
nucleus from triploid endosperm of *Scilla sibirica*. Each filament (loop) cor-
responds to an extended segment in the axis of the three nucleolar chromosomes; the less extended one reflecting reduced gene activity. The thickness is
due largely to DNA, gene product (probably ribosomal RNA) and protein.
Detergent preparation (Schedule 18, La Cour 1966). Phase contrast, × 1500.

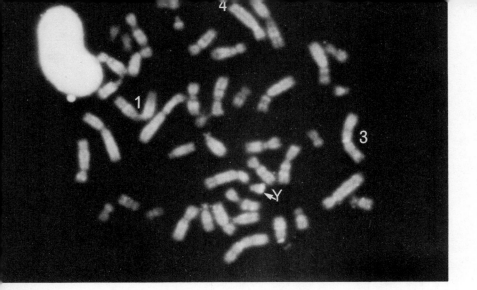

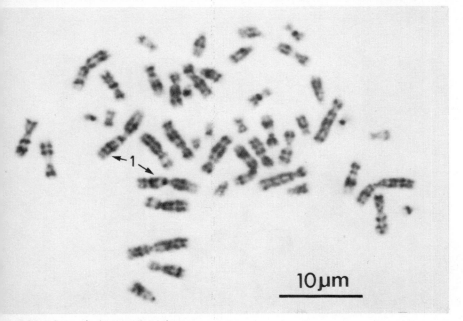

10μm

PLATE XXX

Metaphases in human lymphocytes.

1, From a 46, XY male, stained with 0·5% quinacrine dihydrochloride to show C-banding. Photographed with a Leitz ortholux microscope with mercury lamp and Ploem incidence illumination with BG 12 exciting filter and 510 nm barrier filter. Note intense fluorescence distally in long arm of the Y chromosome and, in this particular individual, at the centromeric regions of chromosomes 3 and 4. Chromosome 1 has been arrowed so that the banding pattern can be compared with that observed using a G-banding technique (see Evans *et al*. 1971, *Chromosoma* **35**: 310). Prep. and photo by Prof. H. J. Evans, × 2150.

2, From a 46, XY male, stained with the ASG technique to show G-banding. The air-dried preparation was incubated for 1 hr at 60° C in 2 × SSC and stained for 1½ hr in 2% Gurr's Giemsa R66, (see Sumner *et al*. 1971, *Nature New Biol*. **232**: 31). Note the similarity in banding pattern obtained with Q- and G-banding techniques. Prep. and photo by Prof. H. J. Evans group, × 2150.

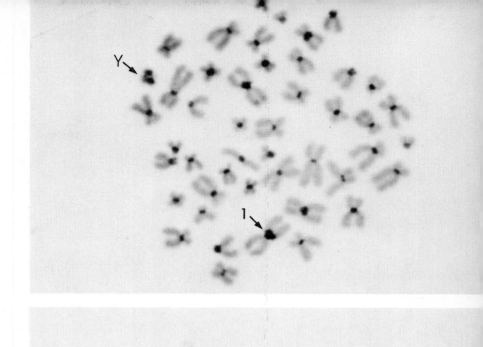

PLATE XXXI

Examples of centromeric heterochromatin in mitotic and meiotic chromosomes.

1, Human lymphocyte metaphase from a 46, XY male heterozygous for a deficiency in the short arm of chromosome 9. The air-dried preparation was C-banded, using a technique including treatment for 1 hr in 0·2 M HCl at room temperature, a 5 min exposure to 5% aqueous barium hydroxide at 50° C and incubation for 1 hr at 60° C in 2 × SSC, followed by staining for 1 hr in 2% Giemsa at pH 6·8 (see Sumner 1972 *Expl. Cell Res.* **75** : 304). Note the very large C-band in chromosome 1 and the dense staining of the Y chromosome. Prep. and photo by Dr A. T. Sumner, × 2150.

2, Metaphase I in a PMC of *Nigella damascena* $2n = 12$, from a squash preparation in which the spread cells were subsequently treated for 20 min in 45% acetic acid at 60° C before proceeding with the G-banding technique (Schedule 18). The heterochromatic regions, which are stained intensely with Giemsa, are confined almost entirely to the centromeres. Prep. and photo by G. E. Marks, × 2000.

Abbreviations

PMC = pollen mother cell.
EMC = embryo-sac mother cell.
SMC = spermatocyte or sperm mother cell.
PT = pollen tube.
GN and VN = generative and vegetative nucleus of PG.
PG = pollen grain.
RT = root tip.
SG = salivary gland (with polytene nuclei).
X and Y = sex chromosomes.
MEIOSIS: The two mitoses by which a diploid mother cell gives four haploid spores or gametes.
I and II = univalents and bivalents at meiosis.
MI, MII and AI, AII, first and second metaphases and anaphases of meiosis.
X, Xta = chiasma, chiasmata, structures produced by crossing over.
CV = crystal violet.
x = basic chromosome number of a polyploid series.
n and $2n$ = gametic and zygotic chromosome numbers.
μm = micrometre = 0·001 mm.
nm = nanometre = 0·0001 μm.
I.E.P. = iso-electric point.
n.a. = numerical aperture.
pH = hydrogen ion concentration.
R = Röntgen unit.
B″ = chromosome break.
B′ = chromatid break.
R′ = chromatid reunion.
AT = adenine: thymine.
GC = guanine: cytosine.
BUdR = 5-bromodeoxyuridine.
IUdR = 5-iododeoxyuridine.
BCdR = 5-bromodeoxycytidine.
2 x (SSC)=0·3 M sodium chloride – 0·03 M trisodium citrate.
U.S.P. units=United States Pharmacopœia units.

Glossary

Apomixis: the occurrence of the external forms of sexual reproduction with the omission of fertilisation and usually meiosis as well. Opposed to true sexual reproduction (*v. Parthenogenesis*).

Asynapsis: non-pairing of chromosomes at prophase or first metaphase of meiosis.

Autosomes: those chromosomes whose segregation does not normally affect the determination of sex. Opposed to sex-chromosomes.

B *Chromosome:* see *Chromosome*.

Basic number: the supposed number of chromosomes found in the gametes of a diploid ancestor of a polyploid; represented by *x* (*v. Chromosome Set*).

Breakage: the breaking of a chromosome (B″) or a chromatid (B′) into two parts which move separately in the ensuing mitosis. It is usually followed during the resting stage by a reunion of chromosomes (R″) or chromatids (R′) or sister reunion (SR).

Cell: a unit in the structure of the animal or plant containing one *nucleus*, or several where these lie in a uniform substratum. Also a body derived from such a unit.

Centromere: a particle or gene in the chromosome thread whose special cycles of repulsion and division determine the anaphase and terminalisation movements of the chromosomes.

    Acentric, Dicentric, Polycentric, Metacentric, Acrocentric, Telocentric: of chromosomes with none, two, or many, or with median or subterminal or terminal centromeres.

    Orientation: the movement of centromeres so that they lie axially with respect to the spindle, either as to their potential halves at metaphase of mitosis (auto-orientation) or as to members of a pair at the first metaphase of meiosis (co-orientation).

    Misdivision: division of the centromere crosswise instead of lengthwise especially at meiosis, to give two telocentric chromosomes.

    Diffuse centromere: where the activity of one or more centromeres is transferred to the ends of the chromosome.

    Neocentric: of chromosomes where the activity of the centromeres is exceptionally transferred to the ends at an abnormal meiosis.

Centrosome: the self-propagating body which divides during cell division in many organisms to form the two poles of the spindle and determine its orientation.

Chiasma: an exchange of partners between pairs of paired chromatids; observed between diplotene and the beginning of the first anaphase in meiosis. See *Crossing Over*.

Interstitial ——: where there is a length of chromatid on both sides of the chiasma.

Terminal ——: where an exchange occurs amongst the end particles of the chromatids, following terminalisation, *q.v.*

Localisation of Chiasmata: the genetic property of restriction of pachytene pairing and chiasma formation to one part of the chromosomes—proximal or distal.

Chimaera: a plant composed of tissues of two genetically distinct types as a result of mutation, segregation, irregularity of mitosis, or artificial fusion by grafting (*v. Mosaic*).

Chromatid: a half chromosome between early prophase and metaphase of mitosis and between diplotene and the second metaphase in meiosis. Afterwards at mitosis it is known as a daughter-chromosome. The separating chromosomes at the first anaphase of meiosis are known as daughter-bivalents, or, if single chromatids derived from the division of univalents, daughter-univalents.

Chromatid Bridge: dicentric chromatid with centromeres passing to opposite poles at anaphase.

Sister Chromatids: those derived from division of one and the same chromosome, as opposed to *non-sister chromatids* which are derived from partner chromosomes at pachytene.

Chromatin: the material in the chromosomes that stains with basic dyes; usually in the resting nucleus. In general DNA *q.v.*

Chromocentre: a condensed body in the resting, interphase or telophase nucleus consisting of fused heterochromatic segments lying near the centromere.

Chromomere: the smallest particle identifiable by its characteristic size and position in the chromosome thread between leptotene and pachytene and in polytene nuclei. Comparable with the gene, *q.v.*

Chromosome: one of the bodies into which the nucleus resolves itself at the beginning of mitosis and from which it is derived at the end of mitosis. (*v.* also *Centromere.*)

Univalent, bivalent, trivalent, multivalent: of bodies representing at meiosis respectively, single unpaired, or two associated, chromosomes or (only in polyploids) three or many associated homologous chromosomes from the mitotic complement.

Chromosome Thread: the thread consisting of centromere, chromomeres, and fine connecting thread at prophase; and constituting, as a spiral, the metaphase chromosome.

Chromosome Set: a minimum complement of chromosomes derived from the gametic complement of a supposed ancestor; *v. Basic Number.*

Ring Chromosome: at mitosis, a chromosome with no ends.

Chromosome Ring: at meiosis, chromosomes associated end to end in a ring, usually by terminal chiasmata; especially applied to diploid interchange heterozygotes where four, six or more chromosomes are so associated.

B *Chromosome:* a supernumerary, *i.e.* one of a type which varies in number as: 0, 1, 2, 3 . . . among individuals within sexual populations of a species.

Isochromosome: chromosome with two identical arms formed by sister-reunion within a terminal centromere following misdivision (*e.g.* attached X in *Drosophila melanogaster*).

Diplochromosome: chromosome which has divided twice during the resting stage without division of its centromere.

Lampbrush Chromosome: appearing at a stage corresponding to diplotene in the ovarian oocyte nuclei of a wide variety of animals. Much extended and with lateral looped projections.

Polytene Chromosome: appearing in gland tissues of Diptera; banded and consisting of high multiples of much extended homologous threads associated in parallel. *v. Endopolyploidy.*

Pro-Chromosomes: small condensed chromosomes appearing in the resting nuclei of certain plant species; strictly not *Chromocentres* or *Heterochromatin.*

Clone: a group of organisms descended by mitosis, *i.e.* vegetatively, from a common ancestor; *v. Individual.*

Coil: v. Spiral.

Complement: the group of chromosomes derived from a particular nucleus in gamete or zygote, composed of one, two or more *Chromosome Sets, q.v.*

Configuration: an association of chromosomes at meiosis, segregating independently of other associations at first anaphase.

Congression: the movement of chromosomes on to the metaphase plate led by the orientation of their centromeres (*q.v.*) at mitosis or meiosis.

Conjugation: the pairing of chromosomes or gametes, or the fusion of pairs of nuclei.

Constriction: an unspiralised segment of fixed position in the metaphase chromosome.

Centric or *primary:* one associated with the centromere.

Nucleolar or *secondary:* one associated with the organisation of the nucleolus.

Co-orientation: v. Centromeres.

Crossing Over: the exchange of corresponding segments between corresponding chromatids of different chromosomes by breakage and reunion following pairing; a process inferred genetically from the reassociation of linked factors in mendelian hybrids and cytologically from the reassociation of parts of chromosomes at *Chiasmata, q.v.*

Cytoplasm: the protoplasm of the cell apart from the nucleus.

Deficiency: loss of a terminal acentric segment of a chromosome.

Differential Segment: one in respect of which two pairing chromosomes differ in a complex or sex heterozygote; by contrast with a *Pairing Segment.*

Deletion: loss of an intercalary acentric segment of a chromosome following breakage.

Diminution: the loss or expulsion of a part of the chromosome complement at mitosis so that a daughter lineage of nuclei is formed lacking this part (as in *Ascaris, Miastor*).

Diploid: (i) the zygotic number of chromosomes ($2n$) as opposed to the gametic or haploid number (n); (ii) an organism having two sets of chromosomes ($2x$) as opposed to organisms having one (haploid), three (triploid) or more sets (x, 3x, etc.).

Disjunction: the separation of daughter chromosomes at anaphase, particularly of the first meiotic division.

Distal: that part of a chromosome which relative to another part is farther from the centromere (as opposed to proximal).

DNA: desoxyribose nucleic acid.

Endomitosis: replication of the chromosomes without division of the nucleus (endoduplication) thus giving *Endopolyploidy* as in *Polytene Nuclei (q.v.).* May be recognised in induced mitosis by the production of diplochromosomes and higher chromatid multiples, or by DNA content in resting nuclei.

Euchromatin: v. Heterochromatin.

Gametes: germ cells which are specialised for fertilisation and cannot normally develop without it.

Gametophyte: the haploid plant which produces the gametes where there is an alternation of generations, as opposed to sporophyte, *q.v.*

Gene: a unit of reproduction and hence of crossing over in the hereditary material. Its change may be inter-genic and structural, or intra-genic and molecular. Comparable with the *Chromomere, q.v.*

Generative Nucleus: a product of division of the primary nucleus in the pollen grain; in Angiosperms (apart from Gramineae) it usually divides into two sperm nuclei in the pollen tube; opposed to the *Vegetative* nucleus which normally does not divide again.

Genetic: a property possessed by an organism by virtue of its heredity (as opposed to its environment).

Genotype: the kind or type of the hereditary properties of an individual organism.

Genotypic Control: the control of chromosome behaviour by the hereditary properties of the organism; opposed to structural control at meiosis in a *Structural Hybrid, q.v.*

Haploid: v. Diploid.

Heterochromatin: whole chromosomes or segments of chromosomes which replicate their DNA late, fail to unravel at telophase and remain condensed at interphase, as opposed to *Euchromatin.*

Heteropycnosis (Positive): of precociously spiralised segments or whole chromosomes at early prophase, opposed to *'Negative'* underspiralised segments, *e.g.* in sex chromosomes (*v. Heterochromatin*).

Heterozygote: a zygote derived from the union of gametes dissimilar in respect of the constitution of their chromosomes or from mutation in a homozygote. Used for narrower categories of hybrid, as: Mendelian, Interchange, Reduplication, Fragmentation, Translocation, Deficiency and Sex Heterozygotes. Opposed to *Homozygote.*

Hybrid: used here for the broader categories of heterozygote as:

Numerical Hybrid: one whose parental gametes differed in respect of the number of chromosomes, *e.g.* a triploid.

Structural Hybrid: one whose parental gametes differed in respect of the structure of their chromosomes.

Individual (Genetic): an organism or group of organisms derived by mitosis from a single fertilised egg (see *Clone, Chimaera*).

Interphase: see *Mitosis.*

Maturation: the formation of gametes or spores by meiosis.

Meiosis: a form of nuclear division (often contrasted with *Mitosis q.v.*) in which the nucleus divides twice and the chromosomes once. The prophase of meiosis is the prophase of the first of the two divisions.

First and Second Divisions: the two cell and nuclear divisions.

Leptotene: the single chromosome threads at the early stage of the prophase of meiosis before *Zygotene*, and by extension the stage itself.

Pachytene: the double thread (and the stage at which it occurs) produced by pairing of the chromosomes in the *Zygotene* stage at prophase of meiosis.

Diplotene: the stage of meiosis which follows division of the chromosomes at the end of the pachytene stage.

Diakinesis: the last stage in the prophase of meiosis—immediately before the disappearance of the nuclear membrane.

Merogony: development of part of an egg with the sperm nucleus, but without the egg nucleus; male parthenogenesis.

Mitosis: the process by which daughter chromosomes are separated into two identical groups accomplishing the division of the nucleus. In contradistinction to meiosis.

> *Prophase:* the stage in mitosis or meiosis from the first appearance of the chromosomes to metaphase.

> *Metaphase:* the stage of mitosis or meiosis in which the centromeres of the chromosomes lie in a plane at right angles to the axis of the spindle and half-way between the poles.

> *Anaphase:* the stage at which daughter centromeres at mitosis or co-oriented centromeres at first division of meiosis move apart.

> *Telophase:* the last stage of mitosis, after anaphase movement of the chromosomes has ceased.

> *Interphase:* period, after re-formation of nuclear membrane at telophase until the following prophase, when most of the chemical components of the cell are synthesised. Applied also to the resting stage between two meiotic divisions but not to the final resting stage in a differentiated cell.

Monosomic: a diploid organism lacking one chromosome of its proper complement (*cf. Trisomic, Tetrasomic*).

Mosaic: in animals, the equivalent of a *Chimaera, q.v.* In plants, a chimaera produced by repeated mutation.

Mother Cell: the cell with a diploid nucleus which by meiosis gives four haploid nuclei, *e.g.* the spore mother-cell (SMC) in Bryophyta and Pteridophyta, the microspore or pollen mother-cell (PMC) and the megaspore or embryo-sac mother-cell (EMC) in flowering plants. The sperm mother-cell is known as the *spermatocyte*, the egg mother cell as the *oöcyte*, in animals. Many other terms, used in the lower plants, are unnecessary in general cytology.

Non-disjunction: cytologically, the failure of separation of paired chromosomes at meiosis and their passage to the same pole; genetically, any results that might be imputed to such an abnormality, although usually arising from the failure of pairing, or from multivalent formation.

Nuclear sap: the fluid which is lost by the chromosomes as they contract during prophase and which fills the space of the nucleus.

Nucleolus: a spherical body of RNA, protein, lipid, etc., enveloping and reflecting the activity of usually a specific segment in a nucleolar chromosome which is active from telophase to the following prophase.

Nucleus: a body whose reproduction, or 'division', is co-ordinated with that of the cell in *Mitosis, q.v.*

Oöcyte: the egg mother cell which becomes the egg by meiosis, when it extrudes two *polar bodies*, or one, as in *Parthenogenesis, q.v.*

Pairing of Chromosomes: active, the coming together of chromosomes at Zygotene, or passive, the continuance of their association at the first metaphase of meiosis.

Pairing Segment: see *Differential Segment.*

Parthenogenesis: a form of apomixis in which the female gamete, haploid or diploid, develops without fertilisation.

Phenotype: the external appearance produced by the reaction of an organism of a given genotype with a given environment.

Polymitosis: occurrence of a series of mitoses supernumerary to normal development and occurring in rapid succession with or without division of the chromosomes (in pollen and in tumours).

Polynemy: the supposed multiple character (x2, 4, 8, 16. . .) of the *Chromosome Thread, q.v.*

Polyploid: an organism with more than two sets of homologous chromosomes. The terms used are triploid, tetraploid, pentaploid, hexaploid, heptaploid, octoploid. Higher multiples are best referred to as 14x, 22x and so on. May be unbalanced (*aneuploid, trisomic, q.v.*), even or odd multiples, functionally diploid (*allo-polyploid*) or multiplied without differentiation between sets (*auto-polyploid*).

Polysomic: (*v. Trisomic*).

Polytene: of nuclei in gland cells of Diptera containing *Polytene Chromosomes* (*q.v.*).

Precocity: the property of the nucleus beginning prophase before the chromosomes have divided; characteristic of meiosis.

RNA: ribose nucleic acid.

Reduction: the halving of the chromosome number at meiosis and, by extension, its genetical concomitant of segregation.

Replication: the synthesis of new DNA from pre-existing DNA as part of nuclear division.

Restitution: the rejoining of broken ends of chromosomes as they were before breakage.

Restitution Nucleus: a single nucleus formed through failure of one of the divisions of meiosis.

Satellite: a small terminal segment of a chromosome separated from the rest by a nucleolar constriction.

Segment: a portion of a chromosome which can conveniently be considered as a unit for a given purpose; *e.g. Differential Segment, q.v.*

Sex Chromosome: one which in the heterozygous sex is mated with a dissimilar homologue; the X and Y *Chromosomes, q.v.*
    Opposed to *Autosome, q.v.*

Sex Heterozygote: an individual of the heterozygous sex, heterozygous for the differential segments of X and Y chromosomes.

Sexual Differentiation: the production by an organism or by related organisms of gametes of two sizes such that the larger can fuse only with the smaller, *i.e.* by fertilisation.

Sexual Reproduction: that which requires meiosis and fertilisation alternating in a cycle.

Sister Reunion: the terminal union of two sister *Chromatids* (*q.v.*) usually following *Breakage* (*q.v.*) of the mother chromosome.

Sperm: the male gamete in animals.

Spermatocyte: the sperm mother-cell.

Spindle: a bipolar molecularly orientated structure, organised by centrosomes or centromeres, or both, within which the centromeres are held in their orientation, co-orientation and separation during metaphase and anaphase.

Spiral: a coiled or helical arrangement of the chromosome thread in a chromosome or chromatid, at mitosis or meiosis.

   Internal Spiral: a coil within a single chromatid between prophase and anaphase.

   Major and Minor Spirals: the larger and smaller of the two internal coils at meiosis.

   Molecular Spirals: the coiling within the chromosome thread which conditions internal spiralisation.

Relic Spiral: the relaxed coiling which survives from the internal coiling of one mitosis at the following telophase and prophase.

Relational Spiral: coiling of two chromatids or chromosomes round one another.

Spiralisation: the assumption of an internal (but not a relational) spiral by the chromatids in mitosis and meiosis.

Spore: a cell specialised for reproduction but not for fertilisation. In the higher plants it is always the product of meiosis and gives rise, after one, two or three mitoses, to gametes (see *Mother Cell*).

    Microspore: a spore which produces a gametophyte bearing only male gametes; a pollen grain.

    Mega- or Macrospore: a spore having the property of giving gametophytes with only female gametes; the embryo-sac.

Sporophyte: the spore-producing diploid phase in the higher plants; *v. Gametophyte.*

Structure (Genetic): the potentially permanent linear order of the particles, chrommeres or genes, in the chromosomes; *v. Hybrid.*

    Structural Change: change in the genetic structure of the chromosome. Of two main types: *interchange,* exchange of segments between two non-homologous chromosomes; *inversion,* reversal of a segment within one chromosome with respect to its own centromere (*v. Hybrid*).

Terminalisation: expansion of the association of the two pairs of chromatids on one side of a chiasma at the expense of that on the other side. So called because the resulting 'movement' of the chiasma is towards the ends of the chromosomes.

Tetrad: spherical quartet of cells formed by meiosis in microspore (pollen) mother cells of plants.

Trisomic, Tetrasomic: of a diploid with one or two extra chromosomes of one type.

X *Chromosome:* (i) with diploid sex differentiation, the sex chromosome in regard to which one sex is homozygous—this is said to be the homozygous sex; (ii) with sex differentiation in the haploid, the sex chromosome of the female.

Y *Chromosome:* the sex chromosome that is present and pairs with the X in the sex heterozygote.

Zygote: the cell formed by the union of gametes and the individual derived from it (*v. Heterozygote*).

BIBLIOGRAPHY*

A. Books and Reviews

Abbreviations: P., Proceedings.   S.T., Stain Technology.

Alfert, M., 1954. Composition and structure of giant chromosomes. *Int. Rev. Cytology* **3**: 131–74. New York (Academic Press).

Allfrey, V. G., Mirsky, A. E., and Stern, H., 1955. The chemistry of the cell nucleus. *Adv. Enzym.* **16**: 411–500.

Anderson, N. G., 1956. Techniques for the mass isolation of cellular components. *Physical Techniques in Biological Research* **3**: 300–52. New York (Academic Press).

Bacq, Z. M., and Alezander, P., 1955. *Fundamentals of Radiobiology*. London (Butterworth Scientific Publications).

Baker, J. R., 1950. *Cytological Technique*. London (Methuen). 3rd edn. pp. 7 + 211.

Baker, J. R., 1958. *Principles of Biological Microtechnique*. London (Methuen). pp. 357.

Barer, R., 1953. *Lecture Notes on the Use of Microscope*. Oxford (Blackwell).

Barer, R., 1955. Phase contrast microscopy. *Research* **8**: 341–50.

Barer, R., 1956. Phase contrast and interference microscopy in cytology. *Physical Techniques in Biological Research* **3**: 30–90. New York (Academic Press).

Becker, W. A., 1938. Struktur und Doppelbrechung der chromosomen. *Arch. exp. Zelf.* **22**: 196–201.

Beermann, W., 1965. Structure and function of interphase chromosomes. In: *Genetics Today* **2**: 375–84. Geerts, S. J. (ed.). London, New York (Pergamon Press).

Belar, K., 1926. Der Formwechsel der Protistenkerne. *Ergebn. Zool.* **6**: 235–652.

Belar, K., 1929. Die Technik der deskriptiven Zoologie. *Meth. wiss. Biol.* **1**: 638–735.

Benirschke, K. and Hsu, T. C. (eds), 1971. *Chromosome Atlas: Fish, Amphibians, Reptiles and Birds*. Berlin (Springer-Verlag). pp. 225.

Brachet, J., 1950. *Chemical Embryology*. New York (Interscience). English translation of 2nd edn. pp. 9 + 523.

Brachet, J., 1957. *Biochemical Cytology*. New York (Academic Press). pp. 516.

Burnham, C. R., 1962. *Discussions in Cytogenetics*. Minneapolis (Burgess Pub. Co.). pp. 366.

Callan, H. G., 1955. Recent work on the structure of cell nuclei. 'Fine Structure of Cells', Noordhof, Groningen.

Caspersson, T., 1950. *Cell Growth and Cell Function*. A cytochemical study. New York (Norton). pp. 185.

* Including some not mentioned in text.

154   Bibliography

Chamot, E. M. and Mason, G. W., 1938. *Handbook of Chemical Microscopy.* New York (Wiley).

Chargaff, E. and Davidson, J. N. (ed.), 1955. *The Nucleic Acids: Chemistry and Biology. New York* (Academic Press).

Conn, H. J., 1953. *Biological Stains.* Geneva, N.Y. (Biotech. Pub.). 6th edn. pp. 9 + 358.

Crick, F. H. C., 1954. Structure and function of DNA. *Discovery* January 1954: 12–17.

Darlington, C. D., 1937. *Recent Advances in Cytology.* London (Churchill). 2nd edn. pp. 16 + 671.

Darlington, C. D. and Upcott, M. B., 1939. The measurement of packing and contraction in chromosomes. *Chromosomes* **1**: 23–32.

Darlington, C. D. and La Cour, L. F., 1945. Chromosome breakage and the nucleic acid cycle. *J. Genet.* **46**: 180–267.

Darlington, C. D. and Wylie, A. P., 1955. *Chromosome Atlas of Flowering Plants.* London (Allen & Unwin).

Davidson, J. N., 1950. *The Biochemistry of the Nucleic Acids.* London (Methuen). 2nd edn. pp. 6 + 194.

Davies, H. G., 1958. The determination of mass and concentration by microscope interferometry. *General Cytochemical Methods* **1**: 57–161. New York (Academic Press).

Demerec, M., 1950. *Biology of Drosophila.* New York (Wiley). p. 632.

Demerec, M. and Kaufmann, B. P., 1961. *Drosophila Guide.* Washington (Carnegie Institute). 7th edn revised. pp. 47.

Dermen, H., 1940. Colchicine polyploidy and technique. *Bot. Rev.* **6**: 599–635.

Drew, K. M., 1955. Life histories in the Algae. With special reference to the Chlorophyta, Phaeophyta and Rhodophyta. *Biol Rev.* **30**: 343–90.

Eberle, P., 1966. Die Chromosomenstruktur des Menschen in Mitosis and Meiosis. Stuttgart (Gustav Fischer Verlag).

Edwards, R. G. and Fowler, R. E., 1970. The genetics of human pre-implantation development. In: *Modern Trends in Human Genetics.* Emery, A. E. H. (ed.). London (Butterworths). pp. 181–213.

Eigsti, O. J. and Dustin, P., 1955. *Colchicine—in Agriculture, Medicine, Biology and Chemistry.* Iowa (Iowa State College Press). p. 470.

Evans, H. J., 1962. Chromosome aberrations induced by ionizing radiations. *Int. Rev. Cytol.* **13**: 221–321.

Ford, C. E., 1962. Human chromosomes. Little club clinics in developmental medicine. No. 5 (Chromosomes in medicine). pp. 48–60.

Ford, E. H. R., 1973. *Human Chromosomes.* London, New York (Academic Press). pp. 381.

Geitler, L., 1938. *Chromosomenbau.* Berlin (Borntraeger). pp. 7 + 190.

Geitler, L., 1953. *Endomitose und endomitotische Polyploidisierug.* Wien, Innsbruck (Springer-Verlag), pp. 89.

Geitler, L., 1955. Normale und Pathologische Anatomie der Zelle. *Handb. Pfl. Physiol.* **1**: 123–67.

Glick, D., Engstrom, A. and Malmstrom, B O G., 1951. A critical evaluation of quantitative histo—and cytochemical microscopic techniques. *Science* **114**: 253–58.

Godward, M.B.E. (ed.), 1966. *The Chromosomes of the Algae.* London (Edward Arnold). pp. 212.

Haddow, A. (ed.), 1952. *Biological Hazards of Atomic Energy.* Oxford (Clarendon Press).

Hale, A. J., 1958. *The Interference Microscope in Biological Research*. Edinburgh (Livingstone). pp. 114.

Hamerton, J. L., 1961. Sex chromatin and human chromosomes. *Int. Rev. Cytol.* **12**: 1–68.

Haskell, G., 1961. Practical Heredity with *Drosophila*. London (Oliver and Boyd). pp. 12 + 124.

Heitz, E. *et al.*, 1956. *Chromosomes*. (Lectures held at the Conference on Chromosomes: Wageningen). Zwolle (Tjeenk Willink).

Hsu, T. C. and Benirschke, K. (ed.), 1970. *An Atlas of Mammalian Chromosomes* Vols. 1–4. Berlin (Springer-Verlag). pp. 800, plates 50.

Imms, A. D., 1938. *A General Textbook of Entomology*. London (Methuen). 4th edn. pp. 12 + 727.

International symposium on the nucleolus—its structure and function. *Nat. Cancer Inst. Monogr.* **23**, 1966.

Jackson, R. C., 1973. Chromosome evolution in *Haplopappus gracilis*: a centric transposition race. *Evolution* **27**: 243–256.

Johansen, D. A., 1940. *Plant Microtechnique*. New York (McGraw-Hill). pp. 531.

John, B. and Lewis, K. R., 1965. The meiotic system. *Protoplasmatologia* **6**: 3–335. Vienna, New York (Springer-Verlag).

Kasten, F. H., 1960. The chemistry of Schiffs reagent, *Int. Rev. Cytol.* **10**: 1–93. New York (Academic Press).

Kihlmann, B. A., 1966. Actions of chemicals on dividing cells. New Jersey (Prentice Hall Inc.) p. 260 et seq.

Knaysi, G., 1951. *Elements of Bacterial Cytology*. Cornell (Comstock Publishing Co. Inc.) 2nd edn.

Koller, P. C., 1957. The genetic component of cancer. *Cancer* **1**: 335–403.

Kurnick, N. B., 1955. Histochemistry of nucleic acids. *Int. Rev. Cytol.* **4**: 221–68.

Lea, D. E., 1955. *Actions of Radiations on Living Cells*. Cambridge (University Press). 2nd edn. pp. 18 + 395.

Lee, Bolles, 1937. *The Microtomist's Vade-Mecum*. Gatenby, J. B. and Painter, T. S. (eds.). London (Churchill). 10th edn. pp. 9 + 784.

Leuchtenberger, C., 1958. Quantitative determination of DNA in cells by Feulgen microspectrophotometry. *General Cytochemical Methods* **1**: 220–78. New York (Academic Press).

Lewitsky, G. A., 1931. The morphology of the chromosomes. *B. appl. Bot.* **27**: 19–137.

Lima-de-Faria, A., 1958. Recent advances in the study of the kinetochore. *Int. Rev. Cytol.* **7**: 123–57. New York (Academic Press).

Lorbeer, G., 1934. Die Zytologie der Lebermoose mit besonderer Berücksichtigung allgemeiner Chromosomenfragen. *Jbn. wiss. Bot.* **80**: 568–817.

McElroy, W. D. and Glass, B. (eds.), 1957. A Symposium on *The Chemical Basis of Heredity*. Baltimore (Johns Hopkins Press).

McLeish, J. and Snoad, B., 1958. *Looking at Chromosomes*. London (Macmillan). pp. 88.

Maheshwari, P., 1950. *An Introduction to the Study of Angiosperms*. New York (McGraw-Hill). pp. 21 + 433.

Makino, S., 1951. *Chromosome Numbers in Animals*. Iowa (Iowa State College Press). 2nd edn. pp. 290.

Manton, I., 1950. *Problems of Cytology and Evolution in the* Pteridophyta. Cambridge (University Press). pp. 6 + 310.

Martin, L. C. and Johnson, E. K., 1931. *Practical Microscopy*. London (Blackie). 1931. pp. 116.

Matsuura, H. and Suto, T., 1935. Contributions to the idiogram study in phanerogamous plants. I. *J.F. Sci. Hokkaido V*, **5**: 33–75.

Michel, K., 1957. *Die Mikrophotographie*. Wien (Springer-Verlag). pp. 10 + 730.

Mitchison, J. M., 1971. *The Biology of the Cell Cycle*. Cambridge (University Press). pp. 313.

Needham, G. H., 1958. *The Use of the Microscope Including Photomicrography*. Springfield, Ill. (Thomas).

Oehlkers, P. F., 1956. *Das Leben Der Gewächse*, Ein Lehrbuch der Botanik, Band I: Die Pflanze als Individuum. Berlin (Springer-Verlag).

Paris Conference, 1971. Standardization in human cytogenetics. Hamerton, J. L., Jacobs, P. A. and Klinger, H. P. (eds.). In: *Birth Defects*—original article series. U.S.A. (The National Foundation).

Pearse, A. G. E., 1953. *Histochemistry—Theoretical and Applied*. London (Churchill). pp. 530.

Pelc, S. R., 1957. Quantitative aspects of autoradiography. *Exp. Cell Res.* (suppl.) **4**: 31–37.

Pelc, S. R., 1958. Autoradiography as a cytochemical method with special reference to C^{14} and S^{35}. *General Cytochemical Methods* **1**: 279–316. New York (Academic Press).

Prescott, D. M. (ed.). *Methods in Cell Physiology*. New York (Academic Press). pp. 465 *et seq*.

Reich, E. and Goldberg, I. H., 1964. Actinomycin and nucleic acid function. In: *Progress in Nucleic Acid Research and Molecular Biology*. Davidson, J. N. and Cohn, W. E. (eds.). New York (Academic Press). pp. 184–230.

Rhoades, M. M. and McClintock, B., 1935. The cytogenetics of maize. *Bot. Rev.* **1**: 292–325.

Rieger, R. and Michaelis, A., 1958. *Genetisches und Cytogenetisches Wörterbuch*. Berlin (Springer-Verlag). 2nd edn.

Riley, R. and Lewis, K. R. (eds.), 1966. *Chromosome Manipulation and Plant Genetics*. Edinburgh (Oliver and Boyd). pp. 123.

Rogers, A. W., 1967. *Techniques of Autoradiography*. London (Elsevier). p. 335 et seq.

Von Rosen, G., 1954. Radiomimetic reactivity. *Socker. Handlingar* II **8**: 157–273.

Schmidt, W. J., 1937. *Die Doppelbrechung von Karyoplasma, Zytoplasma und Metaplasma*. Berlin (Borntraeger). pp. 11 + 388.

Schrader, F. and Hughes-Schrader, S., 1931. Haploidy in Metazoa. *Q. Rev. Biol.* **6**: 411–38.

Sharma, A. K., 1956. Fixation of plant chromosomes. Principles, limitations, recent developments. *Bot. Rev.* **22**: 665–95.

Sharma, A. K. and Sharma, A., 1965. *Chromosome Techniques, Theory and Practice*. London (Butterworths). p. 474 et seq.

Shillaber, C. P., 1944. *Photomicrography*. New York (Wiley).

Smith, D. R. and Davidson, W. M. (eds.), 1958. Symposium on Nuclear Sex. London (Heinemann).

Smith, S. G., 1960. Chromosome numbers of Coleoptera II. *Can. J. Genet. Cytol.* **2**: 66–88.

Sparrow, A. H. and Forro Jr., F., 1953. Cellular radiobiology. *Ann. Rev. Nuclear Sci.* **3**: 339–68.

Sparrow, A. H., Binnington, J. P. and Pond, V., 1958. Bibliography on the effects of ionizing radiations on plants. *Brookhaven National Laboratory Publication*, N.Y. State.

Staniland, L. N., 1952. *The Principles of Line Illustration.* London (Burke). pp. 224.

Stebbins, G. L., 1941. Apomixis in the Angiosperms. *Bot. Rev.* **7**: 507–42.

Steedman, H. F., 1960. *Section Cutting in Microscopy.* Oxford (Blackwell). pp. 7 + 165.

Stevens, G. W., 1957. *Microphotography at Extreme Resolution.* New York (Wiley).

Swift, H., 1953. Quantitative aspects of nuclear nucleoproteins. *Int. Rev. Cytol.* **2**: 1–71. New York (Academic Press).

Swift, H. and Rasch, E., 1956. Microphotometry with visible light. *Physical Techniques in Biological Research* **3**: 354–400. New York (Academic Press).

Taylor, J. H., 1953. Autoradiographic detection of incorporation of P^{32} into chromosomes during meiosis and mitosis. *Exp. Cell. Res.* 4, **1**: 164–73.

Taylor, J. H., 1956. Autoradiography at the cellular level. *Physical Techniques in Biological Research* **3**: 546–76. New York (Academic Press).

Todd, A. R., 1954. The chemistry of the nucleotides. *P.R.S., A.* **226**: 70–82.

Uber, F. M., 1940. Microincineration and ash analysis. *Bot. Rev.* **6**: 204–26.

Vendrely, R. and Vendrely, C., 1956. The results of cytophotometry in the study of the desoxyribonucleic acid (DNA) content of the nucleus. *Int. Rev. Cytol.* **5**: 171–97. New York (Academic Press).

Vincent, W. S., 1955. Structure and chemistry of nucleoli. *Int. Rev. Cytol.* **4**: 269–98. New York (Academic Press).

Visser, T., 1955. Germination and storage of pollen. *Mededlingen van de Landbouwhogeschool te Wageningen/Nederland* **55**: (1) 1–68.

White, M. J. D., 1940. The origin and evolution of multiple sex-chromosome mechanisms. *J. Genet.* **40**: 303–36.

White, M. J. D., 1942. *The Chromosomes.* London (Methuen). 2nd edn. pp. 9 + 124.

White, M. J. D., 1954. *Animal Cytology and Evolution.* Cambridge (University Press). 2nd edn. pp. 18 + 434.

White, M. J. D., 1973. *The Chromosomes.* London (Chapman and Hall). pp. 214.

Zeiger, K., 1938. Physikochemische Grundlagen der histologischen Methodik. *Wiss. Forsch. Ber.* **48**: 1–202.

B. Special Articles

Abe, K., 1933. Mitosen in antheridium von *Sargassum confusum*. *Ag. Sci. Rep. Tohoku* **8**: 259–62.

Adkisson, K. P., Perreault, W. J. and Gay, H., 1971. Differential fluorescent staining of *Drosophila* chromosomes with quinacrine mustard. *Chromosoma* **34**: 190–205.

Alfert, M., 1952. Studies on basophilia of nucleic acids: the methyl green stainability of nucleic acids. *Biol. Bull.* **103**: 145–56.

Alfert, M. and Geschwind, I. I., 1953. A selective staining method for the basic proteins of cell nuclei. *P.N.A.S.* **39**: 991–99.

Allen, C. E., 1935. The occurrence of polyploidy in *Sphaerocarpus*. *Am. J. Bot.* **22**: 635–44.

Allfrey, V. G., Mirsky, A. E. and Stern, H., 1955. The chemistry of the cell nucleus. *Advance. Enzymol.* **16**: 411–500.

Anderson, E. and Sax, K., 1934. A cytological analysis of sterility in *Tradescantia. Bot. Gaz.* **43**: 609–21.

Anderson, E. G., 1935. Chromosomal interchange in maize. *Genetics* **20**: 70–83.

Anderson, R. L., 1936. Effects of temperature on fertilization in *Habrobracon. Genetics* **21**: 467–72.

Andersson, E., 1947. A Case of Asyndesis in *Picea abies. Hereditas* **33**: 302–47.

Ansley, H. R., 1954. A cytological and cytophotometric study of alternative pathways of meiosis in the house centipede (*Scutigera forceps, Rafinesque*). *Chromosoma* **6**, Part 8: 656–95.

Arrighi, F. E. and Hsu, T. C., 1971. Localization of heterochromatin in human chromosomes. *Cytogenetics* **10**: 81–86.

Asana, J. J. and Makino, S., 1935. A comparative study of the chromosomes in the Indian dragonflies. *J.F. Sci. Hokkaido VI*, **4**: 67–86.

Asana, J. J., Makino, S. and Niiyama, H., 1940. Variations in the chromosome number of *Gryllotalpa africana*, etc. *J.F. Sci. Hokkaido VI*, **4**: 59–72.

Auerbach, C. and Robson, J. M., 1946. Chemical production of mutations. *Nature* **157**: 302.

Avanzi, M. G., 1951. Ricerche sulla poliploida somatica nei tessuti differenziati della radico di alcure graminaceae. *Caryologia* **3**: 351–69.

Avanzi, S. and D'Amato, F. D., 1972. Pattern of binding of tritiated actinomycin D to onion chromosomes in fixed material. *Accad. Naz. Lincei, Rend Classe Scienze fis mat. nat. ser VIII* **52**: 215–219.

Babcock, E. B., and Navashin, M., 1930. The genus *Crepis. Bibliogr. Genet.* **6**: 1–90.

Backman, E., 1935. A rapid combined fixing and staining method for plant chromosome counts. *S.T.* **10**: 83–6.

Baird, T. T., 1936. Comparative study of dehydration. *S.T.* **11**: 13–22.

Bajer, A. and Bajer, M. J., 1954. Endosperm, material for study on the physiology of cell division. *Acta Soc. Bot. Polomiae* **23**: 69–98.

Bajer, A., 1955. Living smears from endosperm. *Experientia* **11**: 221.

Bajer, A. and Bajer, M. J., 1956. Ciné-micrographic studies on mitosis in endosperm. II. Chromosome, cytoplasmic and Brownian movements. *Chromosoma* **7**: 558–607.

Bajer, A., 1957. Ciné-micrographic studies on mitosis in endosperm. III. The origin of the mitotic spindle. *Exp. Cell Res.* **13**: 493–502 (1957).

Bajer, A. and Allen, R. D., 1966. Structure and Organization of the living mitotic spindle of *Haemanthus* endosperm. *Science* **151**: 572–574.

Baker, J. R., 1941. A fluid for softening tissues embedded in paraffin wax. *J. Roy. Micr. Soc.* **61**: 75–8.

Baker, J. R., 1946. The histochemical recognition of lipine. *Q.J.M.S.* **87**: 441–70.

Baldwin, J. T., 1939. Chromosomes from leaves. *Science* **90**: 240.

Barber, H. N., 1938. Delayed mitosis and chromatid fusion. *Nature*, **141**: 80.

Barber, H. N., 1939. The rate of movement of chromosomes on the spindle. *Chromosoma* **1**: 33–50.

Barber, H. N., 1940. The suppression of meiosis and the origin of diplochromosomes. *P.R.S., B.* **128**: 170–85.

Barber, H. N., 1941. Chromosome Behaviour in *Uvularia. J. Genet.* **42**: 223, 257.

Barber, H. N., 1942a. Pollen-grain division in the Orchidaceae. *J. Genet.* **43**: 97–103.

Barber, H. N., 1942b. The experimental control of chromosome pairing in *Fritillaria. J. Genet.* **43**: 359–74.

Barber, H. N. and Callan, H. G., 1943. The effects of cold and colchicine on mitosis in the newt. *P.R.S., B.* **131**: 258–71.

Barigozzi, C., 1937. Lo studio degli spodogrammi dei cromosomi. *Comment. pontif. Acad.* **1**: 333–49.

Barton, D. W., 1950. Pachytene morphology of the tomato chromosome complement. *Am. J. Bot.* **37**: 639–43.

Bateman, A. J. and Sinclair, W. K., 1950. Mutations induced in *Drosophila* by ingested phosphorus—32. *Nature* **165**: 117–18.

Battaglia, E., 1957. A simplified Feulgen method using cold hydrolysis. *Caryologia* **9**: 372–3.

Bauer, H., 1931. Die Chromosomen von *Tipula paludosa* Meig. in Eibildung und Spermatogenese. *Z. Zellf.* **14**: 138–93.

Bauer, H., 1932. Die Feulgensche Nuklealfärbung in ihrer Anwendung auf cytologische Untersuchungen. *Z. Zellf.* **15**: 225–47.

Bauer, H., 1933. Mikroskopisch-chemischer Nachweis von Glycogen und einigen anderen Polysacchariden. *Z. mik-anat. Forsch.* **33**: 143–60.

Bauer, H., 1935. Der Afbau der Chromosomen aus den Speicheldrüsen von Chronomus Thummi Kiefer. *Z. Zellf.* **23**: 280–313.

Bauer, H., 1936. Structure and arrangement of salivary gland chromosomes in *Drosophila* species. *P.N.A.S.* **22**: 216–22.

Bauer, H., 1940. Über die Chromosomen der bisexuellen under parthenogenetischen Rasse des Ostracoden, etc. *Chromosoma* **1**: 620–37.

Bauer, H., Demerec, H. and Kaufmann, B. P., 1938. X-ray induced chromosomal alterations in *Drosophila melanogaster*. *Genetics* **23**: 610–30.

Beadle, G. W., 1933. Further studies of asynaptic maize. *Cytologia* **4**: 269–87.

Beams, H. W. and King R. L., 1935. The effect of ultracentrifuging on the cells of the root tip of the bean (*Phaseolus vulgaris*). *P.R.S., B.* **118**: 264–76.

Beatty, A. V., 1937. A method for growing and for making permanent slides of pollen tubes. *S.T.* **12**: 13–14.

Becker, W. A., 1938. Recent investigations *in vivo* on the division of plant cells. *Bot. Rev.* **4**: 446–72.

Beermann, W., 1952. Chromomerenkonstanz und Spezifische Modifikationen der Chromosomenstruktur in der Entwicklung und Organdifferenzierung von *Chironomus tentans*. *Chromosoma* **5**: 139–98.

Beermann, W., 1954. Weibliche Heterogametie bei Copepoden. *Chromosoma* **6**: 381–96.

Beermann, Wolfgang, 1956. Nuclear differentiation and functional morphology of chromosomes. Cold Spring Harbour Symposia on Quantitative Biology. 21.

Bêlanger, L. F. and Le Blond, C. P., 1946. A method for locating radioactive elements in tissues by covering histological sections with a photographic emulsion. *Endocrinology* **39**: 8–13.

Bêlanger, L. F., 1950. A method for routine detection of radio-phosphates and other radioactive compounds in tissues. The inverted autograph. *Anat. Rec.* **107**: 149–56.

Belar, K., 1929*a*. Beiträge zur Kausalanalyse der Mitose II. Untersuchungen an den Spermatocyten von *Chorthippus U.S.W. Arch. EntMech. Org.* **118**: 359–484.

Belar, K., 1929*b*. Beiträge zur Kausalanalyse der Mitose III. Untersuchungen an den Staubfäden, Haarzellen und Blattmeristemzellen von *Tradescantia virginica. Z. Zellf.* **10**: 73–134.

Belar, K., 1933. Zur Teilungsautonomie der Chromosomen. W. Huth (ed.). *Z. Zellf.* **17**: 51–66.

Belling, J., 1926. The iron-acetocarmine method of fixing and staining chromosomes. *Biol. Bull.* **50**: 160–2.

Belling, J., 1928. A method for the study of chromosomes in pollen mother cells. *Univ. Calif. Pubn. Bot.* **14**: 293–9.

Benazzi, M., 1957. Introduzione alla analisi genetica dei biotipi cariologici di tricladi. *Ric. Sci.* **27** (Suppl.): 4–18.

Berenbraum, M. C., 1958. The histochemistry of bound lipids. *Q. J. M. S.* **99**: 231–242.

Bernal, J. D., 1940. Structural units in cellular physiology in 'The cell and protoplasm'. *Amer. Assoc. Adv. Sci. Pub.* **14**: 199–205.

Bernardo, F. F., 1965. Processing *Gossypium* microspores for first division chromosomes. *S.T.* **40**: 205–206.

Bianchi, N., Lima-de-Faria, A., and Jaworska, H., 1964. A technique for removing silver grains and gelatin from tritium autoradiographs of human chromosomes. *Hereditas* **51**: 207–211.

Bishop, C. J., 1950. Differential X-ray sensitivity of *Tradescantia* chromosomes during the mitotic cycle. *Genetics* **35**: 175–87.

Blakeslee, A. F. and Avery, A. G., 1937. Methods of inducing doubling of chromosomes in plants. *J. Herred.* **28**: 392–411.

Bloch, D. P., 1966. Histone differentiation and nuclear activity. *Chromosoma* **19**: 317–339.

Boothroyd, E. R. and Lima-de-Faria, A., 1964. DNA synthesis and differential reactivity in the chromosomes of *Trillium* at low temperature. *Hereditas* **52**: 122–126.

Bosemark, N. O., 1957. Further studies on accessory chromosomes in grasses. *Hereditas* **43**: 236–97.

Bougin, J. P. and Nitsch, J. P., 1967. Obtention de *Nicotiana* haploides à partir de 'etamines cultivées *in vitro*. *Ann Physiol. veg.* **9**: 377–382.

Boveri, T., 1911. Über das Verhalten der Geschlechtschromosomen bei Hermaphroditismus, Beobachtungen an Rhabditis nigrovenosa. *Verh. Phys.-Med. Ges. Würzburg* **41**: 83–97.

Brachet, J., 1940*a*. La détéction histochimique des acides pentosenucléiques. *Compt. Rend. Soc. Biol.* **133**: 88–90.

Brachet, J., 1940*b*. La localisation de l'acide thymonucléique pendant l'oogénèse et la maturation chez les amphibiens. *Compt. Rend. Soc. Biol.* **133**: 90–1.

Brachet, J., 1953. The use of basic dyes and ribonuclease for the cytochemical detection of ribonucleic acid. *Q.J.M.S.* **94**: 1–10.

Bradbury, Q. C., 1931. A new dehydrating agent for histological technique. *Science* **74**: 225.

Breckon, G. and Evans, E. P., 1969. A combined toluidine blue stain and mounting medium. In: *Comparative Mammalian Cytogenetics*. Benirschke, K. (ed.). New York (Springer). pp. 465–466.

Breuer, M. and Pavan, C., 1955. Behaviour of polytene chromosomes of *Rhynchosciara*. *Chromosoma* **7**: 371–86.

Brewbaker, J. L. and Kwack, B. H., 1963. The essential role of calcium ion on pollen germination and pollen tube growth. *Am. J. Bot.* **50**: 859–865.

Brewen, J. G. and Peacock, W. J., 1968. The effect of tritiated thymidine on sister or chromatid exchange in a ring chromosome. *Mutation Res.* **7**: 433–440.

Bridges, C. B., 1935. The vapor method of changing reagents, and of dehydration. *S.T.* **12**: 51–2.

Brieger, F. G. and Graner, E. A., 1943. On the cytology of *Tityus Bahiensis* with special reference to meiotic prophase. *Genetics* **28**: 269–74.

Brink, R. A., 1944. A hybrid between *Hordeum jubatum* and *Secale cereale*. *J. Hered*. **35**: 67–75.

Britten, R. J. and Kohne, D. E., 1968. Repeated sequences in DNA. *Science* **161**: 529–540.

Brock, R. D., 1954. Spontaneous chromosome breakage in *Lilium* endosperm. *Ann. Bot*., **17**: 10–14 (N.S.).

Brown, Robert, 1833. On the organs and mode of fecundation in Orchideae and Asclepiadeae. *Trans. Linn. Soc*. **16**: 685–745.

Brown, Spencer W., 1949. The structure and meiotic behaviour of the differentiated chromosomes of tomato. *Genetics* **34**: 437–61.

Brown, Spencer W., 1954. Mitosis and meiosis in *Luzula campestris* DC. *Univ. Calif. Publ. Bot*. **27**: 231–78.

Buchholz, J. T., 1931. The dissection, staining and mounting of styles in the study of pollen tube distribution. *S.T*. **6**: 13–24.

Buck, J. B. and Boche, R. D., 1938. Some properties of living chromosomes. *Anat. Record* **72**: abstr. 88.

Burch, C. R. and Stock, J. P. P., 1942. Phase-contrast microscopy. *J. Sci. Instr*. **19**: 71.

Butcher, R. L. and Fugo, N. W., 1967. Overripeness and the mammalian ova. II. Delayed ovulation and chromosome abnormalities. *Fertil. Steril*. **18**: 297–302.

Callan, H. G., 1940. The chromosomes of the cymothoid isopod *Anilocra*. *Q.J.M.S*., **82**: 327–35.

Callan, H. G., 1941. The sex-determining mechanism of the earwig. *Forficula auricularia. J. Genet*. **41**: 349–74.

Callan, H. G., 1942. Heterochromatin in *Triton. P.R.S*., *B*. **130**: 324–335.

Callan, H. G. and Spurway, H., 1951. A study of meiosis in interracial hybrids of the newt, *Triturus cristatus. J. Genet*. **50**: 235–49.

Callan, H. G., 1955. Recent work on the structure of cell nuclei. In symposium on fine structure of cells. *Inst. Un. Biol. Sciences* Series *B* **21**: 89. Noordhoff, Groningen.

Callan, H. G. and Lloyd, L., 1956. Visual demonstration of allelic differences within cell nuclei. *Nature* **178**: 355–57.

Callan, H. G., 1957. The lampbrush chromosomes of *Sepia officinalis* and their structural relationship to the lampbrush chromosomes of Amphibia. *Pubbl. Staz. Zool. Napoli* **29**: 329–46.

Callan, H. G., 1966. Chromosomes and nucleoli of the axolotl, *Ambystoma mexicanum. J. Cell. Sci*. **1**: 85–108.

Callan, H. G. and Jacobs, P. A., 1957. The meiotic process in *Mantis religiosa* males. *J. Genet*. **55**: 200–17.

Callan, H. G. and MacGregor, H. C., 1958. Action of deoxyribonuclease on lampbrush chromosomes. *Nature* **181**: 1479–80.

Callan, H. G. and Taylor, J. H., 1968. A radiographic study of the time course of male meiosis in the newt *Triturus vulgaris. J. Cell. Biol*. **3**: 615–26.

Callan, H. G., 1972. Replication of DNA in the chromosomes of eukaryotes. *P.R.S*., *B*. **181**: 19–41.

Camargo, E. P. and Plaut, W., 1967. The radioautographic detection of DNA with tritiated actinomycin D. *J. Cell. Biol*. **35**: 713–716.

Capinpin, J. M., 1930. Brazilin stain on smear preparations of *Oenothera* pollen mother cells. *Science* **72**: 370–1.

Carlson, J. G., 1935. A rapid method for removing cover glasses of microscope slides. *Science* **81**: 365.

Carlson, J. G., 1950. Effects of radiation on mitosis. *J. cell. comp. Physiol.* **35**: (Suppl.) 89–101.

Carlson, J. G., Harrington, N. G. and Gaulden, M. E., 1953. Mitotic effects of prolonged irradiation with low-intensity gamma rays on the *Chortophaga* neuroblast. *Bio. Bull.* **104**: 313–22.

Carlson, J. G. and Harrington, N. G., 1955. X-ray-induced 'stickiness' of the chromosomes of the *Chortophaga* neuroblast in relation to dose and mitotic stage at treatment. *Radiation Res.* **2**: 84–90.

Carpenter, D. C. and Nebel, B. R., 1931. Ruthenium tetroxide as a fixative in cytology. *Science* **74**: 225.

Carr, D. H. and Walker, J. E., 1961. Carbol fuchsin as a stain for human chromosomes. 1961. *S.T.* **36**: 233–236.

Carson, H. L., 1946. The selective elimination of inversion dicentric chromatids during meiosis in the eggs of *Sciara impatiens*. *Genetics* **31**: 95–113.

Caspersson, T., 1939. Über die Rolle der Desoxyribosenukleinsäure bei der Zellteilung. *Chromosoma* **1**: (1) 147–56.

Caspersson, T., 1940. Die Eiweissverteilung in den Strukturen des Zellkerns. *Chromosoma* **1**: (5) 562–604.

Caspersson, T., 1940. Methods for the determination of the absorption spectra of cell structures. *J.R.M.S.* **60**: 8–25.

Caspersson, T., 1941. Einiges über optische Anisotropic und Feinbau von Chromatin und Chromosomen. *Chromosoma* **2**: 247–50.

Caspersson, T., Farber, S., Foley, G. E., Wagh, U. and Zech, L., 1968. Chemical differentiation along metaphase chromosomes. *Exp. Cell Res.* **49**: 214–222.

Caspersson, T., Zech, L., Modest, E. J., Foley, G. E., Wagh, U. and Simonsson, E., 1969a. Chemical differentiation with fluorescent alkylating agents in *Vicia faba* metaphase chromosomes. *Exp. Cell Res.* **58**: 128–140.

Caspersson, T., Zech, D., Modest, E. J., Foley, G. E., Wagh, U. and Simonsson, E., 1969b. DNA – binding fluorochromes for the study of the organization of the metaphase nucleus. *Exp. Cell Res.* **58**: 141–151.

Castro, D., Camara, A. and Malheiros, Nydia, 1949. X-rays in the centromere problem of *Luzula purpurea* Link. *Genet. iber.* **1**: 48–54.

Catcheside, D. G., 1938a. The effect of X-ray dosage upon the frequency of induced structural changes in the chromosomes of *Drosophila melanogaster*. *J. Genet.* **36**: 307–20.

Catcheside, D. G., 1938b. The bearing of the frequencies of X-ray induced interchanges in maize upon the mechanism of their induction. *J. Genet.* **36**; 321–8.

Catcheside, D. G. and Lea, D. E., 1943. The effect of ionization distribution on chromosome breakage by X-rays. *J. Genet.* **45**: 186–96.

Celarier, R. P., 1955. Cytology of the Tradescanteae. *Bull. Torrey Bot. Club.* **82**: 30–38.

Celarier, R. P. and Mehra, K. L., 1958. Determination of polyploidy from herbarium specimens. *Rhodora* **60**: 89–97.

Chapelle, A. de la, Schröder, J. and Selander, R. K., 1971. Repetitious DNA in mammalian chromosomes. *Hereditas* **69**: 149–153.

Chapelle, A. de la, Schröder, J. and Selander, R. K., 1973. *In situ* localization and characterization of different classes of chromosomal DNA: acridine orange and quinacrine mustard fluorescence. *Chromosoma* **40**: 347–360.

Chen, Tse-Tuan, 1936a. Observations on mitosis in Opalinids (Protozoa, Ciliata). I. *P.N.A.S.* **22**: 594–602.

Chen, Tse-Tuan, 1936b. Observations on mitosis in Opalinids (Protozoa, Ciliata). II. *P.N.A.S.* **22**: 602–7.

Chen, Tse-Tuan, 1940. Polyploidy and its origin in *Paramecium*. *J. Hered.* **31**: 175–84.

Chen, Tse-Tuan, 1946. Conjugation in *Paramecium bursaria*. II. Nuclear phenomena in lethal conjugation between varieties. *J. Morph.* **79**: 125–262.

Chen, Tse-Tuan, 1948. Chromosomes in Opalinidae (Protozoa, Ciliata) with special reference to their behaviour, morphology, individuality, diploidy, haploidy, and association with nucleoli. *J. Morph.* **83**: 281–358.

Claude, A., 1941. Particulate components of cytoplasm. *Quant. Biol.* **9**: 263. *Cold Spring Harbour Symp.*

Claude, A. and Potter, J. S., 1943. Isolation of chromatin from the resting nucleus of leukaemic cells. *J. Exp. Med.* **77**: 345–54.

Cleveland, L. R., 1949. The whole life cycle of chromosomes and their coiling systems. *Trans. Am. Phil. Soc.*, **39** (i): 1–100.

Cleveland, L. R., 1949. Hormone-induced sexual cycles of Flagellates. I. Gametogenesis, fertilization, and meiosis in *Trichonympha*. *J. Morph.* **85**: 197–296.

Clowes, F. A. L., 1956. Localization of nucleic acid synthesis in root meristems. *J. exp. Bot.* **7**: 307–12.

Cole, E. C., 1926. A rapid iron haematoxylin technique. *Science* **64**: 452–3.

Coleman, L. C., 1938. Preparation of leucobasic fuchsin for use in the Feulgen reaction. *S.T.* **13**: 123–4.

Coleman, L. C., 1940. The structure of homotypic and somatic chromosomes. *Am. J. Bot.* **27**: 683–88.

Colombo, G., 1953. Eterocromosomi e differenziazione del sesso. Osservazioni Sulle Cellule Germinali di *Anacridium aegyptium* (Acridoidea, Orthoptera). *Riv. Biol.* **46**: 106–13.

Comings, D. E., Avelino, E., Okado, T. A. and Wyandt, H. E., 1973. The mechanism of C- and G-banding of chromosomes. *Exp. Cell Res.* **77**: 469–493.

Conger, A. D. and Fairchild, L. M., 1952. Breakage of chomosomes by oxygen. *Proc. nat. Acad. Sci.* **38**: 289–99.

Conger, A. D. and Fairchild, L. M., 1953. A quick-freeze method for making smear slides permanent. *S.T.* **28**: 281–83.

Conger, A. D., 1953. Culture of pollen tubes for chromosomal analysis at the pollen tube division. *S.T.* **28**: 289–93.

Cooper, K. W., 1937. Reproductive behaviour and haploid parthenogenesis in the grass mite, *Pediculopsis graminum* (Reut.) (Acarina, Tarsonemidae). *P.N.A.S.* **23**: 41–4.

Cullis, C. A. and Schweizer, D., 1974. Repetitious DNA of some *Anemone* species. *Chromosoma* **44**: 417–421.

d'Amato, F., 1950. Differenziazione Istologica per Endopoliploidia Nelle Radici di Alcune Monocotiledoni. *Caryologia* **3**: 12–26.

d'Amato, F., 1952. New evidence on endopolyploidy in differentiated plant tissues. *Caryologia* **4**: 121–44.

d'Amato, F., 1952. Further investigations on the mutagenic activity of acridines. *Caryologia* **4**: 388–413.

d'Amato, F. and d'Amato-Avanzi, M. G., 1954. The chromosome-breaking effect of coumarin derivatives in the *Allium* test. *Caryologia* **6**: 134–50.

Danon, M. and Sachs, L., 1957. Sex chromosomes and human sexual development. *Lancet:* 20–5.

Darlington, C. D., 1936. The external mechanics of the chromosomes. I–V. *P.R.S., B.* **121**: 264–319.

Darlington, C. D., 1939. The genetical and mechanical properties of the sex chromosomes V. *Cimex* and the Heteroptera. *J. Genet.* **39**: 101–37.

Darlington, C. D., 1942. Chromosome chemistry and gene action. *Nature* **149**: 66–69.

Darlington, C. D., Hair, J. B. and Hurcombe, R., 1951. The history of garden hyacinths. *Heredity* **5**: 233–52.

Darlington, C. D. and Haque, A., 1955. The timing of mitosis and meiosis in *Allium ascalonicum*: A problem of differentiation. *Heredity* **9**: 117–27.

Darlington, C. D. and Koller, P. C., 1947. The chemical breakage of chromosomes. *Heredity* **1**: 187–222.

Darlington, C. D. and La Cour, L., 1938. Differential reactivity of the chromosomes. *Ann. Bot.* N.S. **2**: 615–25.

Darlington, C. D. and La Cour, L., 1940. Nucleic acid starvation in chromosomes of *Trillium. J. Genet.* **40**: 185–213.

Darlington, C. D. and La Cour, L., 1941. The detection of inert genes. *J. Hered.* **32**: 114–21.

Darlington, C. D. and La Cour, L. F., 1941. The genetics of embryo-sac development. *Ann Bot.* **5**: 547–62.

Darlington, C. D. and La Cour, L. F., 1945. Chromosome breakage and the nucleic acid cycle. *(Trillium, Tradescantia, Vicia). J. Genet.* **46**: 180–267.

Darlington, C. D. and La Cour, L. F., 1952. The classification of radiation effects at meiosis. *Heredity* **6**: 41–55.

Darlington, C. D. and Osterstock, H. C., 1936. Projection method for demonstration of chromosomes *in situ. Nature* **138**: 79.

Darlington, C. D. and Shaw, G. W., 1959. Parallel polymorphism in the heterochromatin of *Trillium* species. *Heredity* **13**: 100–30.

Darlington, C. D. and Thomas, P. T., 1941. Morbid mitosis and the activity of inert chromosomes in *Sorghum. P.R.S., B.* **130**: 127–50.

Darlington, C. D. and Upcott, M. B., 1941a. The activity of inert chromosomes in *Zea mays. J. Genet.* **41**: 275–96.

Darlington, C. D. and Upcott, M. B., 1941b. Spontaneous chromosome change. *J. Genet.* **41**: 297–338.

Das, N. K., 1965. The inactivation of the nucleolar apparatus during neiotic prophase in corn anthers. *Exp. Cell Res.* **40**: 360–4.

Das, N. K. and Alfert, M., 1963. Silver staining of a nucleolar fraction, its origin and fate during the mitotic cycle. *Ann. Histochim.* **8**: 109–114.

Davidson, D., 1957. The irradiation of dividing cells. I. The effects of X-rays on prophase chromosomes. *Chromosoma* **9**: 39–60.

Davidson, D., 1958a. The irradiation of dividing cells: II. Changes in sensitivity to X-rays during mitosis. *Ann. Bot., N.S.* **22**: 183–195.

Davidson, D., 1958. The effects of chelating agents on cell division. *Exp. Cell Res.* **14**: 329–32.

Davidson, J. N. and Waymouth, C., 1944. The histochemical demonstration of ribonucleic acid in mammalian liver. *P.R.S. Edin.* **62**: 96–8.

Davies, H. G., Wilkins, M. H. F., Chayen, J. and La Cour, L. F., 1954. The use of the interference microscope to determine dry mass in living cells and as a quantitative method. *Q.J.M.S.* **95**: 271–304.

Day, P. R., Boone, D. M. and Keitt, G. W., 1956. *Venturia inaequalis* (CKE.) Wint. XI, The chromosome number. *Am. J. Bot.* **43**: 835–38.

Deeley, E. M., 1955. An integrating microdensitometer for biological cells. *J. Sci. Instr.* **32**: 263–67.

Deeley, E. M., Richards, B. M., Walker, P. M. B. and Davies, H. G., 1954. Measurements of Feulgen stain during the cell-cycle with a new photoelectric scanning device. *Exp. Cell Res.* **6**: 569–72.

de Lamater, E. D., 1951. A staining and dehydrating procedure for the handling of micro-organisms. *S.T.* **26**: 199–204.

Dermen, H., 1941. Intranuclear polyploidy in bean induced by naphthalene acetic acid. *J. Hered.* **32**: 133–38.

de Tomasi, J. A., 1936. Improving the technique of the Feulgen stain. *S.T.* **11**: 137–44.

Dewey, W. C. and Humphrey, R. M., 1965. Increase in radiosensitivity to ionizing radiation related to replacement of thymidine in mammalian cells with 5-bromodeoxyuridine. *Radiation Res.* **26**: 538.

Dobzhansky, T., 1934. Studies on hybrid sterility. I. Spermatogenesis in pure and hybrid *Drosophila pseudoobscura*. *Z. Zellf.* **21**: 169–223.

Doniach, I. and Pelc, S. R., 1950. Autoradiograph technique. *Brit. J. Radiol.* **23**: 184–92.

Doroshenko, A. V., 1928. Pollen physiology (in Russian). *Bull. appl. Bot.* **18**: (5) 217–344.

Dorsey, E., 1936. Induced polyploidy in wheat and rye. *J. Hered.* **27**: 154–60.

Dowrick, G. J., 1952. The chromosomes of *Chrysanthemum*. I. The species. *Heredity* **6**: 365–75.

Dowrick, G. J., 1957. The influence of temperature on meiosis. Heredity **11**: 37.

Drew, K. M., 1939. An investigation of *Plumaria elegans*, etc. *Ann. Bot.* N.S. **3**: 347–67.

Dufrenoy, J., 1935. A method for embedding plant tissues without dehydration. *Science* **82**: 235.

Duryee, W. R., 1937. Isolation of nuclei and non-mitotic chromosome pairs from frog eggs. *Arch, exp. Zellforsch.* **19**: 171.

Duryee, W. R., 1950. Chromosomal physiology in relation to nuclear structure. *Ann. N.Y. Acad. Sci.* **50**: 920–53.

Dutrillaux, B., 1973. Application to the normal karyotype of R-band and G-band techniques involving proteolytic digestion. In: *Chromosome Identification* – technique and applications in biology and medicine. Caspersson, T. and Zech, L. (ed.). *P. 23rd Nobel Symposium*, pp. 38–42. Stockholm (The Nobel Foundation). New York, London (Academic Press).

Ebstein, B. S., 1967. Tritiated actinomycin – D as a cytochemical label for small amounts of DNA. *J. Cell. Biol.* **35**: 709–713.

Ehrenberg, L. and Gustafsson, A., 1957. On the mutagenic action of ethylene oxide and diepoxybutane in barley. *Hereditas* **43**: 595–602.

Ellenhorn, J., 1933. Experimental-photographische Studien der lebenden Zelle. *Z. Zellf.* **20**: 288–308.

Ellenhorn, J., *et al.*, 1935. The optical dissociation of *Drosophila* chromomeres by means of ultra-violet light. *C.R. Acad. Sci. URSS* 1935, **1** (4) 234–42.

Elliott, C. G., 1955. The effect of temperature on chiasma frequency. *Heredity* **9**: 385–98.

Elliott, C. G., 1956. Chromosomes in micro-organisms. VI. *Symp. Soc. gen. Microbiol.* pp. 279–95.

Emerson, S., 1938. The genetics of incompatibility in *Oenothera organensis*. *Genetics* **23**: 190–202.

Emsweller, S. L., 1944. Improving smear techniques by the use of enzymes. *S.T.* **19**: 109–14.

Emsweller, S. L. and Jones, H. A., 1935. Meiosis in *Allium fistulosum, Allium cepa* and their hybrid. *Hilgardia* **9**: 277–288.

Endicott, K. M. and Jagoda, H., 1947. Microscopic historadiographic technique for locating and quantitating radioactive elements in tissue. *P. Soc. exp. Biol. N.Y.* **64**: 170–72.

Erickson, R. O., Sax, K. O. and Ogar, M., 1949. Perchloric acid in the cyto-chemistry of pentose nucleic acid. *Science* **110**: 472–73.

Ernst-Schwarzenbach, M., 1957. Zur Kenntnis der Fortpflanzungsmodi der Braunalge *Halopteris filicina*. *Pubbl. Staz. Zool. Napoli*. **29**: 347–88.

Evans, E. P., Burtenshaw, M. D. and Ford, C. E., 1972. Chromosomes of mouse embryos and new-born young: preparations from membranes and tail tips. *S.T.* **47**: 229–234.

Evans, H. J., 1964. Uptake of ^{3}H-thymidine and patterns of DNA replication in nuclei and chromosomes of *Vicia faba*. *Exp. Cell Res*. **35**: 381–393.

Evans, H. J., 1972. Properties of human X and Y sperm. In: The genetics of spermatozoon. Beatty, R. A. and Glueckson-Waelsch, S. (ed.). *P. Int. Symp*. Edinburgh and New York Departments of Genetics of the University of Edinburgh and the Albert Einstein College of Medicine: 144–159.

Evans, H. J., 1972. Molecular architecture of human chromosomes. *Brit. Med. Bull.* **29** No. 3: 196–202.

Evans, H. J., Buckton, K. E. and Sumner, A. T., 1971. Cytological mapping of human chromosomes: results obtained with quinacrine fluorescence and the acetic–saline techniques. *Chromosoma* **35**: 310–325.

Evans, T. C., 1947. Radioautographs in which the tissue is mounted directly on the photographic plate. *P. Soc. exp. Biol. N.Y.* **64**: 313–15.

Evans, W. L., 1956. The effect of cold treatment on the desoxyribonucleic acid (DNA) content in the cells of selected plants and animals. *Cytologia* **21**: 417–32.

Fabergé, A. C., 1940. An experiment on chromosome fragmentation in *Tradescantia* by X-rays. *J. Genet.* **39**: 229–48.

Fabergé, A. C., 1945. Snail stomach cytase, a new reagent for plant cytology. *S.T.* **20**: 1–4.

Fabergé, A. C., 1955. The analysis of induced chromosome aberrations by maize endosperm phenotypes. *Z. Vererbungslehre*. **87**: 392–420.

Fabergé, A. C. and La Cour, L., 1936. An electrically heated needle for paraffin embedding. *Science* **84**: 142.

Fahmy, O. G. and Fahmy, M. J., 1956. Cytogenetic analysis of the action of carcinogens and tumour inhibitors in *Drosophila melanogaster* V. *J. Genet.* **54**: 146–64.

Fankhauser, G., 1937a. The production and development of haploid salamander larvae. *J. Hered.* **28**: 1–15.

Fankhauser, G., 1937b. The development of fragments of the fertilised *Triton* egg with the egg nucleus alone ('gyno-merogony'). *J. exp. Zool.* **75**: 413–70.

Fankhauser, G., 1941. The frequency of polyploidy and other spontaneous aberrations of chromosome number among larvae of the newt, *Triturus viridescens*. *P. N. A. S.* **27**: 507–12.

Fankhauser, G., 1945. The effects of changes in chromosome number on amphibian development. *Quarterly Rev. Biol.* **20**: 20–78.

Fankhauser, G., 1948. The organization of the amphibian egg during fertilization and cleavage. *Ann. New York Acad. Sci.* **49**: 684–708.

Fankhauser, G. and Humphrey, R. R., 1952. The rare occurrence of mitosis without spindle apparatus ('colchicine mitosis') producing endopolyploidy in embryos of the Axolotl. *Proc. Nat. Acad. Sci.* **38**: 1073–82.

Favorsky, M. V., 1939. New polyploidy-inducing chemicals. *C.R. Acad. Sci. URSS* **25**: 71–4.

Fernandes, A., 1939. Sur le comportement d'un chromosome surnuméraire pendant la mitose. *Sci. genet.* **1**: 141–67.

Feulgen, R. and Imhauser, K., 1925. Über die für Nuklealreaktion und Nukleal-färbung verantwortlich zu machenden Gruppen, etc. II. *Hoppe-Seyl Z.* **148**: 1–16.

Feulgen, R. and Rossenbeck, H., 1924. Mikroscopisch-chemischer Nachweis einer Nucleinsäure vom Typus der Thymonucleinsäure. *Hoppe-Seyl Z.* **135**: 203–48.

Flax, M. H. and Himes, M. H., 1952. A microspectrophotometric analysis of metachromatic staining of nucleic acids in tissues. *Physiol. Zool.* **25**: 297–311.

Foot, K. and Strobell, E. C., 1905. Prophases and metaphases of the first maturation spindle of *Allolobophora foetida. Am. J. Anat.* **4**: 199–243.

Ford, C. E. and Evans, E. P., 1969. Meiotic preparations from mammalian testes. In: *Comparative Mammalian Cytogenetics*. Benirschke, K. (ed.). New York (Springer). pp. 461–464.

Ford, C. E. and Hamerton, J. L., 1956. A colchicine, hypotonic citrate squash sequence for mammalian chromosomes. *S.T.* **31**: 247–251.

Ford, C. E., Hamerton, J. L. and Sharman, G. B., 1957. Chromosome poly-morphism in the common shrew. *Nature* **180**: 392–93.

Ford, C. E., Jacobs, P. A. and Lajtha, L. G., 1958. Human somatic chromosomes. *Nature* **181**: 1565–68.

Ford, E. H. R. and Woollam, D. H. M., 1963. A colchicine, hypotonic citrate, air drying sequence for mammalian chromosomes. *S.T.* **38**: 271–274.

Fraser, H. C. I., 1908. Contributions to the cytology of *Humaria rutilans*, Fries. *Ann. Bot.* **22**: 35–55.

Gall, J. G., 1952. The lampbrush chromosomes of *Triturus viridescens. Exp. Cell Res. (Suppl.)* **2**: 95.

Gall, J. G., 1955. Structure and function in the amphibian oocyte nucleus. *S.E.B. Symp.* **9**: 358–70.

Gall, J. G. and Callan, H. G., 1962. H³-uridine incorporation in lampbrush chromosomes. *P.N.A.S.* **48**: 562–570.

Gall, J. G. and Pardue, M. L., 1969. Formation and detection of RNA–DNA hybrid molecules in cytological preparations. *P.N.A.S.* **63**: 378–383.

Gallagher, A., Hewitt, G. M. and Gibson, I., 1973. Differential Giemsa staining of heterochromatic B-chromosomes in *Myrmeleotettix maculatus*. (Orthoptera: Acrididae). *Chromosoma* **40**: 167–72.

Ganner, E. and Evans, H. J., 1971. The relationship between patterns of DNA replication and of quinacrine fluorescence of the human chromosome complement. *Chromosoma* **35**: 326–341.

Geard, C. R. and Peacock, W. J., 1969. Sister chromatid exchanges in *Vicia faba. Mutation Res.* **7**: 215–223.

Geitler, L., 1935a. Der Spiralbau somatischer Chromosomen. *Z. Zellf.* **23**: 514–21.

Geitler, L., 1936. Über den feineren Kern-und Chromosomenbau der Clado-phoraceen. *Planta* **25**: 530–78.

Gelei, J., 1921. Weitere Studien über die Oogenese des *Dendrocoelum lacteum*. II. *Arch. Zellf.* **16**: 88–169.

Geyer-Duszynska, I. 1954. X-ray induced fragmentation of salivary gland chromosomes in *Drosophila melanogaster*. *Zool. Polon.* **6**: 251–82.

Geyer-Duszynska, I., 1955. X-ray induced fragmentation of salivary gland chromosomes in *Drosophila melanogaster*. *Zool. Polon.* **6**: 250–81.

Ghosh, C., 1955. Retention of activity by solutions of pancreatic deoxyribonuclease after freezing and thawing. *S.T.* **31**: 17–20.

Giles Jr., H. and Bolomey, R. A., 1948. Cytogenetical effects of internal radiations from radioisotopes. *Symp. Quant. Biol.* **13**: 104–12.

Giles, N., 1940. Spontaneous chromosome aberrations in *Tradescantia*. *Genetics* **25**: 69–87.

Giles, N. H., 1943. Comparative studies of the cytogenetical effects of neutrons and X-rays. *Genetics* **28**: 398–418.

Giles, N. H. and Riley, H. P., 1950. Studies on the mechanism of the oxygen effect on the radiosensitivity of *Tradescantia chromosomes*. *P.N.A.S.* **36**: 337–44.

Giles, N. H., Beatty, A. V. and Riley, H. P., 1952. The effect of oxygen on the production by fast neutrons of chromosomal aberrations in *Tradescantia* microspores. *Genetics* **37**: 641–49.

Godward, M. B. E., 1948. The iron alum acetocarmine method for algae. *Nature* **161**: 203.

Godward, M. B. E., 1954. The 'diffuse' centromere or polycentric chromosomes in *Spirogyra*. *Ann. Bot. Lond. N.S.* **18**: 144–56.

Gottschalk, W., 1955. Die Paarung Homologer Bivalente und der Ablauf von Partnerwechseln in den Frühen Stadien der Meiosis Autopolyploider Pflanzen. *Zeitschr. f. ind. Abst.-u. Vererbgsl.* **87**: 1–24.

Graupner, V. H. and Weisberger, A., 1933. Die Verwendung von Lösungen in Dioxan als Fixierungsmittel für Gefrierschnitte. *Zool. Anz.* **102**: 39–44.

Grell, K. G., 1952. Der Stand unserer Kenntnisse über den Bau der Protistenkerne. *Verh. dtsch. zool. Ges.* (1952). 213–51.

Grell, K. G., 1953. Die Chromosomen von *Aulacantha scolymantha*. *Arch. Protistenk.* **99**: 2–54.

Grell, K. G., 1954. Der Generationswechsel der polythalamen *Rotaliella heterocaryotica*. *Arch. Protistenk.* **100**: 269–86.

Groat, R. A., 1939. Two new mounting media superior to canada balsam and gum damar. *Anat. Rec.* **74**: 1 and Suppl. 1.

Gulick, A., 1941. The chemistry of chromosomes. *Bot. Rev.* **7**: 433–57.

Gustafsson, A., 1937. The different stability of chromosomes and the nature of mitosis. *Hereditas* **22**: 281–335.

Guyénot, E. and Danon, M., 1953. Chromosomes et ovocytes de Batraciens. *Rev. suisse Zool.* **60**: 1.

Haga, T., 1956. Genome and polyploidy in the genus *Trillium*. VI. Hybridization and speciation by chromosome doubling in nature. *Heredity* **10**: 85–98.

Haga, T. and Kurabayashi, M., 1954. Chromosomal variation in natural populations of *Trillium kàmtschaticum*. *Mem. Fac. Sci. Kyushu Univ. Series E.* **1**: 159–85.

Hair, J. B., 1956. Subsexual reproduction in *Agropyron*. *Heredity* **10**: 129–60.

Hance, R. T., 1933. Improving the staining action of iron haematoxylin. *Science* **77**: 287.

Hance, R. T., 1937. Air conditioning for microtomes. *Science* **86**: 313.

Haque, A., 1953. The irradiation of meiosis in *Tradescantia*. *Heredity* **6**: (Suppl.) 57–75.

Haque, A., 1963. Differential labelling of *Trillium* chromosomes by H³ – thymidine at low temperature. *Heredity* **18**: 129–133.

Harland, S. C., 1936. Haploids in polyembryonic seeds of Sea Island cotton. *J. Hered.* **27**: 229–31.

Harland, S. C., 1940. New polyploids in cotton by the use of colchicine. *Trop. Agric. Trin.* **17**: 53–4.

Haupt, A. W., 1935. A gelatin fixative for paraffin sections. *S.T.* **5**: 97–8.

Hayden, B. and Smith, L., 1949. The relation of atmosphere to the biological effects of X-rays. *Genetics* **34**: 26–43.

Hearne, E. M. and Huskins, C. L., 1935. Chromosome pairing in *Melanoplus femur-rubrum*. *Cytologia* **6**: 123–47.

Heitz, E., 1929. Heterochromatin. Chromozentren, Chromomeren. *Ber. dt. bot. Ges.* **47**: 274–286.

Heitz, E., 1931. Die Urasche der gesetzmässigen Zahl, Lage, Form und Grösse pflanzlicher Nucleolen. *Planta* **12**: 774–844.

Heitz, E., 1932. Die Herkunft der Chromozentren. *Planta* **18**: 571–636.

Heitz, E., 1935. Chromosomenstruktur und Gene. *Z.I.A.V.* **70**: 402–47.

Heitz, E., 1936*b*. Die Nucleal-Quetschmethode. *Ber d. bot. Ges.* **53**: 870–8.

Heitz, E. and Bauer, H., 1933. Beweise für die Chromosomennatur der Kernschleifen in Knäuelkernen von Bibio U.S.W. *Z. Zellf.* **17**: 67–82.

Helwig, E. R., 1938. The frequency of reciprocal translocations in irradiated germ cells of *Circotettix verruculatus* (Orthoptera). *Arch. Biol. Paris* **49**: 143–58.

Henderson, S. A., 1964. RNA synthesis during male meiosis. *Chromosoma* **15**: 345–366.

Henderson, S. A., 1970. The time and place of meiotic crossing-over. *Ann. Rev. Genet,* **4**: 295–324.

Henderson, S. A., Nicklas, R. B. and Koch, C. A., 1970. Temperature-induced orientation instability during meiosis: an experimental analysis. *J. Cell Sci.* **6**: 323–350.

Heuser, M. and Razari, L., 1970. A standardized method of peripheral blood culture by cold temperature treatment. In: *Methods in Cell Physiology IV*. Prescott, D. M. (ed.). New York, London (Academic Press). pp. 478–95.

Hillary, B. B., 1939*a*. Use of the Feulgen reaction in cytology. 1. Effect of fixatives on the reaction. *Bot. Gaz.* **101**: 276–300.

Hillary, B. B., 1939*b*. Improvements in the permanent root tip squash technique. *S.T.* **14**: 97–9.

Hillary, B. B., 1940. Uses of the Feulgen reaction in cytology. II. *Bot. Gaz.* **102**: 225–35.

Howard, A. and Pelc, S. R., 1953. Synthesis of DNA in normal and irradiated cells. *Heredity* (Suppl. Chr. Breakage) **6**: 261–73.

Hsu, T.C., 1952. Mammalian chromosomes *in vitro*. The karyotype of man. *J. Hered.* **43**: 167–72.

Hsu, T. C., 1954. Mammalian chromosomes *in vitro*. IV. Some human neoplasms. *J. nat. Cancer Inst.* **14**: 905–33.

Hsu, T. C. and Arrighi, F. E., 1971. Distribution of constitutive heterochromatin in mammalian chromosomes. *Chromosoma* **34**: 243–253.

Hsu, T. C. and Somers, C. E., 1961. Effect of 5-bromodeoxyuridine on mammalian chromosomes. *P.N.A.S.* **47**: 396–403.

Hughes-Schrader, S., 1955. The chromosomes of the giant scale *Aspido-proctus maximus* Louns (Coccoideamargarodidae). With special reference to asynapsis and sperm formation. *Chromosoma* 7: 420–438.

Hungerford, D. A., 1965. Leucocytes cultured from small inocula of whole blood and the preparation of metaphase chromosomes by treatment with hypotonic KCl. *S.T.* 40: 333–38.

Hungerford, D. A., Mellman, W. J. *et al.*, 1970. Chromosome structure in man III. Pachytene analysis and identification of the supernumerary chromosome in a case of Down's syndrome (Mongolism). *P.N.A.S.* 67: 221–24.

Huskins, C. L. and Smith, S. G., 1935. Meiotic chromosome structure in *Trillium erectum* L. *Ann. Bot.* 49: 119–50.

Huskins, C. L. and Steinitz, L., 1948. The nucleus in differentiation and development. II. Induced mitosis in differentiated tissues of *Rhoeo* roots. *J. Heredity* 39: 67–77.

Husted, L. and Burch, P. R., 1946. The chromosomes of polygrid snails. *Amer. Nat.* 80: 410–29.

Husted, L. and Ruebush, T. K., 1940. A comparative cytological and morphological study of Mesostoma, etc. *J. Morph.* 67: 387–410.

Ikeda, K. and Makino, S., 1936. Studies on the sex and chromosomes of the oriental human blood fluke, *Schistosomum japonicum* Kats. *J.F. Sci. Hokkaido VI,* 5: 11–71.

Ikushima, T. and Wolff, S., 1974. Sister chromatid exchanges induced by light flashes to 5-bromodeoxyuridine and 5-iododeoxyuridine-substituted Chinese hamster chromosomes. *Exp. Cell Res.* 87: 15–19.

Iwanami, Y. and Nakamura, N., 1972. Storage in an organic solvent as a means for preserving viability of pollen grains. *S.T.* 47: 137–39.

Jachimsky, H., 1935. Beitrag zur Kenntnis von Geschlechtschromosomen und Heterochromatin bei Moosen. *Jb. wiss. Bot.* 81: 203–38.

Jacobson, W. and Webb, M., 1952. The two types of nucleoproteins and mitosis. *Exp. Cell Res.* 3: 163–83.

Jain, H. K., 1957. Effect of high temperature on meiosis in *Lolium*: nucleolar inactivation. *Heredity* 11: 23–36.

Jensen, C. J., 1964. Pollen storage under vacuum. *Pog. Vet. Agric. Coll.* 1964 year book; 133–46.

Johansen, D. A., 1935. Dehydration and infiltration. *Science* 82: 253–4.

John, B., 1957 XY Segregation in the crane fly *Tipula maxima. Heredity* 11: 209–15.

John, B. and Henderson, S. A., 1962. Asynapsis and polyploidy in *Schistocerca paranensis. Chromosoma* 13: 111–47.

John, B. and Hewitt, G. M., 1965. The B-chromosome system of *Myrmeleo-tettix maculatus* (Thumb.) I. The mechanics. *Chromosoma* 16: 548–78.

John, B. and Hewitt, G. M., 1965. The B-chromosome system of *Myrmeleotettix maculatus* (Thumb.) II. The statics. *Chromosoma* 17: 121–38.

John, B. and Lewis, K. R., 1957. Studies on *Periplaneta americana.* I. Experimental analysis of male meiosis. II. Interchange heterozygosity in isolated populations. *Heredity* 11: 1–22.

Karling, J. S., 1928. Nuclear and cell division in the antheridial filaments of the Characeae. *B. Torrey bot. Cl.* 55: 11–39.

Karpechenko, G. D., 1927. Polyploid hybrids of *Raphanus sativus x Brassica oleracea. B. Appl. Bot.* 17: (3), 305–410.

Karpechenko, G. D., 1938. New tetraploid barleys—the hulled and the naked. *C. R. Acad. Sci. URSS* 21: 59–62.

Karpechenko, G. D., 1940. Tetraploid six-rowed barleys obtained by colchicine treatment. *C.R. Acad. Sci. URSS* **27**: 47–50.

Katayama, H., 1939. The sex chromosomes of a may-fly, *Ameletus costalis* Mats. (Ephemerida). *Jap, J. Genet.* **15**: 139–44.

Kato, H. and Moriwaki, K., 1972. Factors involved in the production of banded structures in mammalian chromosomes. *Chromosoma* **38**: 105–20.

Kaufmann, B. P., 1938. Nucleolus-organizing regions in salivary gland chromosomes of *Drosophila melanogaster. Z. Zellf.* **28**: 1–11.

Kaufmann, B. P., 1946. Organization of the chromosome. I. Break distribution and chromosome recombination in *Drosophila melanogaster. J. exp. Zool.* **102**: 293–320.

Kaufmann, B. P., 1951. Chromosome aberrations induced in animal cells by ionizing radiations. *Radiation Biology* **6**: 27–711.

Kaufmann, B. P. and Gay, H., 1947. The influence of X-rays and near infrared rays on recessive lethals in *Drosophila melanogaster. P.N.A.S.* **33**: 366–72.

Kaufmann, B. P., Hollaender, A. and Gay, H., 1946. Modification of the frequency of chromosomal re-arrangements induced by X-rays in *Drosophila*. I. Use of near infrared radiation. *Genetics* **31**: 349–67.

Kaufmann, B. P., McDonald, M. R. and Gay, H., 1948. The enzymatic degradation of ribonucleoproteins of chromosomes, nucleoli and cytoplasm. *Nature* **162**: 814–15.

Kaufmann, B. P., McDonald, M. R. and Gay, H., 1951. The distribution and interrelation of nucleic acids in fixed cells as shown by enzymatic hydrolysis. *J. cell. comp. Physiol.* **38**: (Suppl.) **1**: 71–99.

Kawaguchi, E., 1938. Der Einfluss der Eierbehandlung mit Zentrifugierung auf die Vererbung bei dem Seidenspinner. II. *Cytologia* **9**: 38–54.

Kelley, E. G., 1939. Reaction of dyes with cell substances IV. Quantitative comparison of tissue nuclei and extracted nucleoproteins. *J. Biol. Chem.* **127**: 55–71.

Keyl, H. G. and Pelling, C., 1963. Differentialle DNS—Replikation in den Speicheldrüsenchromosomen von *Chironomus thummi. Chromosoma* (Berlin) **14**: 347–59.

Kho, Y. O. and Baer, J., 1968. Observing pollen tubes by means of fluorescence. *Euphytica* **17**: 298–302.

Kihara, H., 1919. On the relation between the germination of pollens and the absorption of water. *Sapporo Nat. Hist. Soc.* **7**: 179–84.

Kihara, H., 1924. Cytologische und genetische Studien bei wichtigen Getreidearten. *Mem. Coll. Sci. Kyoto B*, **1**: 1–200.

Kihara, H., 1927*b*. Über die Vorbehandlung einiger pflanzlicher Objecte bei der Fixierung der Pollenmutterzellen. *Bot. Mag. Tokyo* **41**: 124–28.

Kihara, H., 1940. Formation of haploids by means of delayed pollination in *Triticum monococcum. Bot. Mag. Tokyo* **54**: 178–85.

Kihara, H. and Yamashita, K., 1938. Künstliche Erzeugung haploider und triploider Einkornweizen durch Bestäubung mit Röntgenbestrahlten Pollen. *Akemine Commem. Papers* 9–20.

Kihlman, B. A., 1952. Induction of chromosome changes with purine derivatives. *Symb. Bot. Upsal.* **11**: (4). 1–96.

Kihlman, B. A., 1955. Oxygen and the production of chromosome aberrations by chemical and X-rays. *Hereditas* **41**: 384–404.

Kihlman, B. A. and Levan, H., 1951. Localized chromosome breakage in *Vicia faba. Hereditas* **37**: 382–87.

King, D. T., Harris, J. E. and Tkaceyk, S., 1951. Liquid-emulsion methods in electron track radiography. *Nature* **167**: 273.

Kirby-Smith and Daniels, D. S., 1953. The relative effects of X-rays, gamma rays and beta rays on chromosomal breakage in *Tradescantia*. *Genetics* **38**: 375–88.

Klinger, H. P. and Ludwig, K. S., 1957. A universal stain for the sex chromatin body. *S.T.* **32**: 235–44.

Klingstedt, H., 1931. Digametie beim Weibchen der Trichoptere *Limnophilus decipiens* Kol. *Acta zool. fenn.* **10**: 1–69.

Knapp, E., 1935. Zur Frage der genetischen Aktivität des Heterochromatins, nach Untersuchungen am X-Chromosom von *Sphaerocarpus Donnellii*. *Ber. d. Bot. Ges.* **53**: 751–60.

Köhler, A. and Loos, W., 1941. Das Phasenkontrastverfahren und seine Anwendungen in der Mikroskopie. *Naturwiss.* **29**: 49–61.

Koller, P. C., 1935. The internal mechanics of the chromosomes, IV. Pairing and coiling in salivary gland nuclei of *Drosophila*. *P.R.S., B.* **810**: 371–97.

Koller, P. C., 1936. The origin and behaviour of chiasmata, XI. *Dasyurus* and *Sarcophilus*. *Cytologia* **7**: 82–103.

Koller, P. C., 1937. The genetical and mechanical properties of sex chromosomes. III. Man. *P.R.S. Edin.* **57**: 194–214.

Koller, P. C., 1940. Technique for staining mitotic chromosomes of *Drosophila*. *Dros. Inf. Service* **13**: 79–80.

Koller, P. C., 1942. A new technique for mitosis in tumours. *Nature* **149**: 193.

Koller, P. C., 1944. Segmental interchange in mice. *Genetics* **29**: 247–63.

Koller, P. C., 1946. Nucleic acid control in the X chromosomes of the hamster. *P.R.S., B.* **133**: 313–26.

Koller, P. C., 1953. The cytological effects of irradiation at low intensity. *Heredity* (Suppl.) **6**: 5–22.

Koller, P. C., 1956. Cytological variability in human carcinomatosis. *Ann. N.Y. Acad. Sci.* **63**: 793–816.

Koltzoff, N. K., 1938. The structure of the chromosomes and their participation in cell-metabolism. *Biol. Zhur.* **7**: 3–46.

Kostoff, D., 1938*a*. Polyploid plants produced by colchicine and acenapthene. *Curr. Sc.* **7**: 108–10.

Kostoff, D., 1938*b*. The effect of centrifuging upon the germinated seeds of various plants. *Cytologia* **8**: 420–42.

Kostoff, D., 1938*c*. Irregular meiosis and abnormal pollen-tube growth induced by acenaphthene. *Curr. Sc.* **7**: 8–11.

Kotval, J. P. and Gray, L. H., 1947. Structural changes produced in microspores of *Tradescantia* by alpha-radiation. *J. Genet.* **48**: 135–54.

Kunitz, M., 1940. Crystalline ribonuclease. *J. Gen. Physiol.* **24**: 15–32.

Kunitz, M., 1948. Isolation of crystalline desoxyribonuclease from beef pancreas. *Science* **108**: 19–20.

Kunitz, M., 1950. Crystalline desoxyribonuclease. I. Isolation and general properties. *J. Gen. Physiol.* **33**: 349–62.

Kurabayashi, M., 1953. Effects of post-temperature treatments upon the X-ray induced chromosomal aberrations. *Cytologia* **18**: 253–65.

Kurabayashi, M., 1957. Evolution and variation in *Trillium*. IV. Chromosomal variation in natural populations of *Trillium kamtschaticum*. *Jap. J. Bot.* **16**: 1–45.

Kurnick, N. B., 1949. The quantitative estimation of desoxyribose nucleic acid based on methyl-green staining. *Exp. Cell Res.* **1**: 151–58.

Kurnick, N. B., 1955. Pyronin Y in the methyl-green-pyronin histological stain. *S.T.* **30**: 213–30.

Kuwada, Y., 1937. The hydration and dehydration phenomena in mitosis. *Cytologia, Fujii Jub. Vol.* 389–402.

Kuwada, Y. and Nakamura, T., 1934*a*. Behaviour of chromonemata in mitosis, II. *Cytologia* **5**: 244–7.

Kuwada, Y. and Nakamura, T., 1940. Behaviour of cromonemata in mitosis. IX. On the configurations assumed by the spiralised chromonemata. *Cytologia* **10**: 492–515.

Kuwada, Y., Shinke, N. and Oura, G. 1938. Artificial uncoiling of the chromonema spirals as a method of investigation of the chromosome structure. *Z. wiss. Mikr.* **55**: 8–16.

Kwack, B. H. and Kim, I. H., 1967. An improved method for culturing *Tradescantia* pollen tubes for chromosomal analysis. *Cytologia* **32**: 1.

La Cour, L. F., 1931. Improvements in everyday technique in plant cytology. *J.R.M.S.* **51**: 119–26.

La Cour, L. F., 1935. Technique for studying chromosome structure. *S.T.* **10**: 57–9.

La Cour, L. F., 1937. Improvements in plant cytological technique. *Bot. Rev.* **5**: 241–58.

La Cour, L. F., 1941. Acetic-orcein. *S.T.* **16**: 169–74.

La Cour, L. F., 1944. Mitosis and cell differentiation in the blood. *P.R.S. Edin.* **62**: 73–85.

La Cour, L. F., 1949. Nuclear differentiation in the pollen grain. *Heredity* **3**: 319–37.

La Cour, L. F., 1951. Heterochromatin and the organisation of nucleoli in plants. *Heredity* **5**: 37–50.

La Cour, L. F., 1953. The physiology of chromosome breakage and reunion in *Hyacinthus. Heredity* **6**: (Suppl.) 163–79.

La Cour, L. F., 1953*a*. The *Luzula* system analysed by X-rays. *Heredity* **6**: (Suppl.) 77–81.

La Cour, L. F., 1953*b*. The physiology of chromosome breakage and reunion in *Hyacinthus. Heredity* **6**: (Suppl.) Symp. Chr. Breakage, 163–79.

La Cour, L. F., 1954. Smear and squash techniques in plant cytology. *Laboratory Practice* **3**: 326–30.

La Cour, L. F., 1960. The reaction of heterochromatic segments of chromosomes to the Feulgen stain. *John Innes Report* 38–9.

La Cour, L. F., 1963. Ribose nucleic acid and the metaphase chromosome. *Expl. Cell Res.* **29**: 112–118.

La Cour, L. F., 1966. The internal structure of nucleoli. In: *Chromosomes Today.* **1**: 150–60. Darlington, C. D. and Lewis, K. R. (eds.). Edinburgh (Oliver and Boyd).

La Cour, L. F. and Chayen, J., 1958. A cyclic staining behaviour of the chromosomes during mitosis and meiosis. *Exp. Cell Res.* **14**: 462–68.

La Cour, L. F., Chayen, J. and Gahan, P. S., 1958. Evidence for lipid material in chromosomes. *Exp. Cell Res.* **14**: 469–85.

La Cour, L. F., Deeley, E. M., and Chayen, J. 1956. Variations in the amount of Feulgen stain in nuclei of plants grown at different temperatures. *Nature* **177**: 272–73.

La Cour, L. F., and Fabergé, A. C., 1943. The use of cellophane in pollen tube technique. *S.T.* **18**: 196.

La Cour, L. F. and Pelc, S. R., 1958. Effect of colchicine on the utilization of labelled thymidine during chromosomal reproduction. *Nature* **182**: 506–508.

La Cour, L. F. and Rutishauser, A., 1954. X-ray breakage experiments with endosperm. *Chromosoma* **6**: 696–709.

La Cour, L. F. and Wells, B., 1970. Chromocentres and the synaptinemal complex. *J. Cell Sci.* **6**: 655–667.

La Cour, L. F. and Wells, B., 1973. Fine structure and staining behaviour of heterochromatic segments in two plants. *J. Cell Sci.* **14**: 505–21.

Lajtha, L. G., 1959. The culture of bone marrow cells *in vitro. British Med. Bull.* **15**: 47–9.

Lajtha, L. G., 1960. Bone marrow culture. *Methods in Medical Research* **8**: 12–26. Chicago (Yearbook Publishing Co.).

Lane, G. R., 1951. X-ray fractionation and chromosome breakage. *Heredity* **5**: 1–35.

Lantz, L. A. and Callan, H. G., 1954. Phenotypes and spermatogenesis of interspecific hybrids between *Triturus cristatus* and *T. marmoratus. J. Genet.* **52**: 165–85.

Larter, L. N. H., 1932. Chromosome variation and behaviour in *Ranunculus* L. *J. Genet.* **26**: 255–83.

Latt, S. A., 1973. Microfluorometric detection of deoxyribonucleic acid replication in human metaphase chromosomes. *P.N.A.S.* **70**: 3395–99.

Lawson, C. A., 1936. A chromosome study of the aphid *Macrosiphum solanifolii. Biol. Bull.* **70**: 288–307.

Leedale, G. F., 1958. Mitosis of chromosome numbers in the Eugleninae (Flagellata). *Nature* **181**: 502–3.

Levan, A., 1932. Cytology studies in *Allium* II. *Hereditas* **16**: 257–94.

Levan, A., 1936. Die Zytologie von *Allium cepa* × *fistulosum. Hereditas* **21**: 195–214.

Levan, A., 1938. The effect of colchicine on root mitoses in *Allium. Hereditas* **24**: 471–86.

Levan, A., 1939*a*. The effect of colchicine on meiosis in *Allium. Hereditas* **25**: 9–26.

Levan, A., 1939*b*. Amphibivalent formation in *Allium cernuum* and its consequences in the pollen. *Bot. Notiser* **1939**: 236–60.

Levan, A., 1939*c*. Tetraploidy and octoploidy induced by colchicine in diploid *Petunia. Hereditas* **25**: 109–31.

Levan, A., 1939*d*. Cytological phenomena connected with the root swelling caused by growth substance. *Hereditas* **25**: 87–96.

Levan, A., 1940. Meiosis of *Allium porrum*, a tetraploid species with chiasma localisation. *Hereditas* **26**: 454–62.

Levan, A., 1951. Chemically induced chromosome reactions in *Allium cepa* and *Vicia faba. Symp. Quant. Biol.* **16**: 233–43.

Levan, A. and Hauschka, T. S., 1952. Chromosome numbers of three mouse ascites tumours. *Hereditas* **38**: 251–55.

Levan, A. and Hauschka, T. S., 1953. Endomitotic reduplication mechanisms in ascites tumours of the mouse. *J. Nat. Cancer Inst.* **14**: 1–43.

Levan, A. and Lotfy, T., 1949. Naphthalene acetic acid in the *Allium* test. *Hereditas* **35**: 337–74.

Levan, H. and Östergren, G., 1943. The mechanism of C – mitotic action. *Hereditas* **29**: 381–443.

Levi, H., 1954. Quantitative β-track autoradiography of single cells. *Exp. Cell Res.* **7**: 44.

Lewis, D., 1942. The physiology of incompatibility in plants. I. The temperature effect. *P.R.S., B.* **131**: 13–26.

Lewis, D., 1943. The incompatability sieve for producing polyploids. *J. Genet.* **45**: 261–4.

Lewis, D. and Modlibowska, I., 1942. Genetical studies in pears, IV. Pollen tube growth and incompatibility. *J. Genet.* **43**: 211–22.

Lewis, K. R., 1957. Squash techniques in the cytological investigation of mosses. *Trans. Brit. Bryol. Soc.* **3**: 279–84.

Lewitsky, G. A., 1931. An essay on cytological analysis of the fixing action of the chrom-acetic formalin and the chromic formalin. *B. Appl. Bot.* **27**: (1) 176–85.

Lewitsky, G. A., 1934. Fixation changes of the chromosome body. *C.R. Acad. Sci. URSS* **4**: 223–4.

Lewitsky, G. A., 1940. The progeny of X-rayed *Crepis capillaris. Cytologia* **11**: 1–29.

Lilly, L. J., 1958. Effects of cyanide and ionising radiation on the roots of *Vicia faba. Exp. Cell Res.* **14**: 257–67.

Lilly, L. J. and Thoday, J. M., 1956. Effects of cyanide on the roots of *Vicia faba. Nature* **177**: 338–39.

Lima-de-Faria, A., 1952. Chromomere analysis of the chromosome complement of rye. *Chromosoma* **5**: 1–68.

Lima-de-Faria, A., 1959. Differential uptake of tritiated thymidine into hetero- and euchromatin in *Melanoplus* and *Secale. J. Biophys. Biochem. Cytol.* **6**: 457–466.

Lima-de-Faria, A. and Bose, S., 1954. Spectrophotometric analysis of aceto-carmine solutions. *Hereditas* **40**: 419–24.

Lin, T. P., 1954. The chromosomal cycle in *Parascaris equorum (Ascaris megalocephala)*: oogenesis and diminution. *Chromosoma* **6**: 175–98.

Linnert, G., 1950. Die einwirkung von chemikalien auf die meiosis. *Z. Vererbungslehre* **83**: 422–28.

Linskens, H. F. and Esser, K., 1957. Uber eine spezifische Anfärburg der Pollenschläuche ein Griffel und die zall der Kallosepfropfen nach Selbstung und Freindung. *Naturwiss.* **44**: 16.

Lucas, F. F. and Stark, M. B., 1931. A study of living sperm cells of certain grasshoppers by means of the ultraviolet microscope. *J. Morph.* **52**: 91–107.

Lutkov, A. N., 1938. Tetraploidy in *Linum* induced by high-temperature treatment of the zygote. *C.R. Acad. Sci. URSS* **19**: 87–90.

McClintock, B., 1929. A method for making acetocarmine smears permanent. *S.T.* **4**: 53–6.

McClintock, B., 1931. Cytological observations involving known genes, translocations and an inversion in *Zea mays. B.Univ. Mo. Agr. Exp. Sta.* **163**: 1–30.

McClintock, B., 1933. The association of non-homologous parts of chromosomes in the mid-prophase of meiosis *Zea mays. Z. Zellf.* **19**: 191–237.

McClintock, B., 1934. The relation of a particular chromosomal element to the development of the nucleoli in *Zea mays. Z. Zellf.* **21**: 294–328.

McClintock, B., 1945. The chromosomes of *Neurospora crassa. Am. J. Bot.* **32**: 671–8.

McDonald, M. R., 1948. A method for the preparation of 'protease-free' crystalline ribonuclease. *J. gen. Physiol.* **32**: 39–42.

McDonald, M. R. and Kaufmann, B. P., 1954. The degradation by ribonuclease of substrates other than ribonucleic acid. *J. Histochem. & Cytochem.* **2**: 387–93.

176 Bibliography

McDonald, M. R. and Kaufmann, B. P., 1957. Production of mitotic abnormalities by ethylene-diametetraacetic acid. *Exp. Cell Res.* **12**: 415–17.

MacGregor, H. C. and Kezer, J., 1971. The chromosomal localization of a heavy satellite DNA in the testis of *Plethodon C. cinereus. Chromosoma* **33**: 167–182.

McLeish, J., 1953. The action of maleic hydrazide in *Vicia. Heredity* **6**: (Suppl.) 125–47.

McLeish, J., 1954. The consequences of localized chromosome breakage. *Heredity* **8**: 385–407.

McLeish, J., Bell, L. G. E., Là Cour, L. F. and Chayen, J., 1957. The quantitative cytochemical estimation of arginine. *Exp. Cell Res.* **14**: 120–25.

McLeish, J. and Sherratt, H. S. A., 1958. The use of the Sakaguchi reaction for the cytochemical determination of combined arginine. *Exp. Cell Res.* **14**: 625–28.

McLeish, J. and Sunderland, N., 1961. Measurements of deoxyribosenucleic acid (DNA) in higher plants by Feulgen photometry and chemical methods. *Exp. Cell Res.* **24**: 527–40.

McWhorter, F. P., 1939. Application of fine grain processing and condenser illumination enlarging to photo-micrography. *S.T.* **14**: 87–96.

Madge, M., 1936. The use of agar in embedding small or slender objects. *Ann. Bot.* **50**: 677.

Maeda, T., 1939. Chiasma studies in the silkworm, *Bombyx mori* L. *Jap. J. Genet.* **15**: 118–27.

Maheshwari, P., 1939. Recent advances in microtechnique II. The paraffin method in plant cytology. *Cytologia* **10**: 257–81.

Maio, J. and Schildkraut, C. L., 1967. Isolated mammalian metaphase chromosomes. 1. General characteristics of nucleic acids and proteins. *J. Mol. Biol.* **24**: 29–39.

Makino, S., 1939. The chromosomes of the carp, *Cyprinus carpio*, etc. *Cytologia* **9**: 430–40.

Makino, S., 1941. Studies on the murine chromosomes I. *J.F. Sci. Hokkaido VI Zoo.* **7**: 305–80.

Makino, S. and Nakanishi, Y. H., 1955. A quantitative study on anaphase movement of chromosomes in living grasshopper spermatocytes. *Chromosome* **7**: 439–50.

Makino, S., Udagawa, T. and Yamashina, Y., 1956. Karyotype studies in birds. 2: A comparative study of chromosomes in the Columbidae. *Caryologia* **8**: 275–93.

Malheiros, Nydia, Castro, D. and Camara, A., 1947. Cromosomas sem centromero localizado Ocasso da *Luzula purpurea* Link. *Agron. lusit.* **13**: 1–10.

Manton, I., 1939. Evidence on spiral structure and chromosome pairing in *Osmunda regalis* L. *Phil. Trand. R.S.* **230**: 179–215.

Marks, G. E., 1957. The cytology of *Oxalis dispar* (Brown). *Chromosoma* **8**: 650–70.

Marks, G. E., 1973. Arapid HCl/toluidine blue squash technique for plant chromosomes. *S.T.* **48**: 229–231.

Marks, G. E. and Schweizer, D., 1973. Giemsa banding: karyotype differences in some species of *Anemone* and in *Hepatica nobilis. Chromosoma* **44**: 405–416.

Martin, F. M., 1959. Staining and observing pollen tubes in the style by means of fluorescence. *S.T.* **34**: 125–128.

Mather, K., 1932. Chromosome variation in *Crocus* I. *J. Genet.* **26**: 129–42.

Mather, K., 1933. The behaviour of meiotic chromosomes after X-irradiation. *Hereditas* **19**: 303–22.

Mather, K., 1935. Meiosis in *Lilium. Cytologia* **6**: 354–80.

Mather, K. and Stone, L. H. A., 1933. The effect of X-radiation upon somatic chromosomes. *J. Genet.* **28**: 1–24

Matsuura, H., 1937. Chromosome studies on *Trillium kamtschaticum* Pall. V. *Cytologia Fujii Jub. Vol.* 20–34.

Matthey, R., 1941. Etude biologique et cyt. de *Saga pedo* Pallas (Orthoptères-Tettigoniidae). *Rev. suisse zool.* **48**: 91–142.

Matthey, R. and Brink, J. M. Van, 1956. La question des hétérchromosomes chez les Sauropsidés. I. Reptiles. *Experientia* **12**: 53.

Mazia, D., 1954. The particulate organization of the chromosomes. *P.N.A.S.* **40**: 521–27.

Metz, C. W., 1935. Structure of the salivary gland chromosomes in *Sciara*. *J. Hered.* **26**: 177–88.

Meyer, J. R., 1943. Colchicine Feulgen leaf smears. *S.T.* **18**: 53–56.

Meyer, J. R., 1945. Prefixing with paradichlorobenzene to facilitate chromosome study. *S.T.* **20**: 121–25.

Michel, K., 1941. Die Darstellung von Chromosomen mittels des Phasenkontrastverfahrens. *Naturwiss.* **29**: 61–62.

Miller, O. L., 1964. Extrachromosomal nucleolar DNA in amphibian oocytes. *J. Cell Biol.* **23**: 60A.

Milovidov, P. F., 1936. Zür Theorie und Technik der Nukleafärbung. *Protoplasma* **25**: 570–97.

Mirsky, A. E. and Ris, H., 1947. Isolated chromosomes. The chemical composition of isolated chromosomes. *J. gen. Physiol.* **31**: 1–18.

Mirsky, A. E. and Ris, H., 1951. The desoxyribonucleic acid content of animal cells and its evolutionary significance. *J. gen. Physiol.* **34**: 451–62.

Mirsky, A. E. and Ris, H., 1951. Composition and structure of isolated chromosomes. *J. gen. Physiol.* **34**: 475–92.

Moore, K. L., Graham, M. A. and Barr, K. L., 1953. The detection of chromosomal sex in hermaphrodites from skin biopsy. *Surg. Gynec. Obstet.* **96**: 641.

Moorhead, P. S., Nowell, P. W., Mellman, W. J., Battips, D. M., and Hungerford, D. A., 1960. Chromosome preparations of leucocytes cultured from human peripheral blood. *Exp. Cell Res.* **20**: 613–16.

Morrison, J. W., 1953. Chromosome behaviour in wheat monosomics. *Heredity* **7**: 203–17.

Muldal, S., 1952. The chromosomes of the earthworms. I. The evolution of polyploidy. *Heredity* **6**: 55–76.

Muller, H. J., 1927. Artificial transmutation of the gene. *Science* **66**: 84–87.

Muller, H. J., 1940. An analysis of the process of structural change in chromosomes of *Drosophila. J. Genet.* **40**: 1–66.

Müntzing, A., 1941. Differential response to X-ray treatment of diploid and tetraploid barley. *K. Fysiog. Sallsk. Lund. Forh.* **11**: 1–10

Müntzing, A. and Prakken, R., 1941. Chromosomal aberrations in rye populations. *Hereditas* **27**: 273–308.

Nagao, S., 1933. Number and behaviour of chromosomes in the genus *Narcissus. Mem. Coll. Sci. Kyoto* B. **8**: 81–100.

Nakamura, T., 1937. Double refraction of the chromosomes in paraffin sections. *Cytologia Fujii Jub. Vol.* 482–93.

Narasimham, R., 1963. Mass culture of pollen on cellophane – filter paper supports. *S.T.* **38**: 340–341.

Narbel, M., 1946. La cytologie de la parthénogenèse chez *Apterona helix (Lepid. Psychides). Rev. suisse Zool.* **53**: 625–81.

Navashin, M., 1926. Variabilität des Zellkerns bei *Crepis*-Arten in Bezug auf die Artbildung. *Z. Zellf.* **4**: 171–215.

Navashin, M., 1934. Die Methodik der zytologischen Untersuchungen für züchterische Zwecke. *Z. Zucht A.* **19**: 366–413.

Naville, A. and Beaumont, J. de, 1933. Recherches sur les chromosomes des Neuroptères. *Arch. Anat. micr.* **29**: 119–243.

Nebel, B. R., 1931. Lacmoid-martius yellow for staining pollen-tubes in the style. *S.T.* **6**: 27–9.

Nebel, B. R., 1932. Chromosome structure in the Tradescanteae I. *Z. Zellf.* **16**: 252–84.

Nebel, B. R., 1939. Chromosome structure. *Bot. Rev.* **5**: 563–626.

Nebel, B. R., 1940. Chlorazol Black E as an aceto-carmine auxiliary stain. *S.T.* **15**: 69–72.

Nebel, B. R. and Ruttle, M. L., 1939. Colchicine and its place in fruit breeding. *Circ. N.Y. Agr. Exp. Sta.* **183**: p. 19.

Newcombe, H. B., 1942. The action of X-rays on the cell I & II. *J. Genet.* **43**: 145–71, 237–48.

Newcomer, E. H., 1938. A procedure for growing, staining and making permanent slides of pollen tubes. *S.T.* **13**: 89–91.

Newton, W. C. F., 1927. Chromosome studies in Tulipa and some related genera. *J. Linn. Soc. (Bot.)* **47**: 336–54.

Nicklas, R. B. and Staehly, C. A., 1967. Chromosome manipulation. I. The mechanics of chromosome attachment to the spindle. *Chromosoma* **21**: 1–16.

Nicklas, R. B., 1967. Chromosome manipulation. II. Induced reorientation and the experimental control of segregation in meiosis. *Chromosoma* **21**: 17–50.

Nitsch, J. P. and Nitsch, C., 1969. Haploid plants from pollen grains. *Science* **163**: 85–87.

Nogusa, Sh., 1955. Chromosome studies in Pisces, IV. The chromosomes of *Mogrunda obscura* (Gobiidae), with evidence of male heterogamety. *Cytologia* **20**: 11–18.

Nomarski, G., 1955. Microinterféromètre différentiell a ondes polarisées. *J. de Physique et le Radium* **16**: 9–13.

Nowell, P. C., 1960. Phytohemagglutinin: an initiator of mitosis in cultures of normal human leucocytes. *Cancer Research* **20**: 462–6.

Oehlkers, F., 1953. Chromosome breaks influenced by chemicals. *Heredity* **6**: (Suppl.) 95–105.

Ogawa, K., 1954. Chromosome studies in the Myriapoda. VII. A chain-association of the multiple sex-chromosomes found in *Otocryptops sexspinosus. Cytologia* **19**: 265–72.

Ohnuki, Y., 1968. Structure of chromosomes. I. Morphological studies of the spiral structure of human somatic chromosomes. *Chromosoma* **25**: 402–428.

Ojima, Y., 1958. A cytological study on the development and maturation of the parthenogenetic and sexual eggs of *Daphnia pulex* (Crustacea-Cladocera). *Kwansei Univ. Ann. Stud.* **6**: 123–76.

Oksala, T., 1945. Zytologische Studien an Odonaten. III. Die Ovogenese. *Ann Acad. Sc. Fenn, A.* IV **9**: 6–29.

Oksala, T., 1956. The mitotic mechanism of two mouse ascites tumours. *Hereditas* **42**: 161–88.

Olivieri, G. and Brewen, J. G., 1966. Evidence for non-random rejoining of chromatid breaks and its relation to the origin of sister chromatid exchanges. *Mutation Res.* **3**: 237–248.

O'Mara, J. G., 1939. Observations on the immediate effects of colchicine. *J. Heredity* **30**: 35–7.

O'Mara, J. G., 1948. Acetic acid methods for chromosome studies at prophase and metaphase in meristems. *S.T.* **23**: 201–4.

Orian, A. J. E. and Callan, H. G., 1957. Chromosome numbers of Gammarids. *J. mar. biol. Ass.* **36**: 129–42.

Östergren, G., 1954. Polyploids and aneuploids of *Crepis capillaris* produced by treatment with nitrous oxide. *Genetica* **27**: 54–64.

Östergren, G. and Bajer, A., 1958. Permanent preparations from endosperm cells flattened in the living state. *Hereditas* **44**: 466–70.

Östergren, G. and Wakonig, T., 1954. True or apparent sub-chromatid breakage and the induction of labile states in cytological chromosome loci. *Botaniska Notiser* **4**: 357–75.

Oura, G., 1936. A new method of unravelling the chromonema spirals. *Z. wiss. Mikr.* **53**: 36–7.

Pachmann, U. and Rigler, R., 1972. A quantum yield of acridine interaction with DNA of definite base sequences. A basis for explanation of acridine bands in chromosomes. *Exp. Cell Res.* **72**: 602–608.

Painter, T. S., 1934. A new method for the study of chromosome aberrations, etc. *Genetics* **19**: 175–88.

Painter, T. S., 1943. Cell growth and nucleic acids in the pollen of *Rhoeo discolor. Bot. Gaz.* **105**: 58.

Painter, T. S. and Taylor, A. N., 1942. Nucleic acid storage in the toad's egg. *P.N.A.S.* **28**: 311–17.

Pardue, M. L. and Gall, J. G., 1970. Chromosomal localization of mouse satellite DNA. *Science* **168**: 1356–1358.

Partanen, C. R., Sussex, I. M. and Steeves, T. A., 1955. Nuclear behaviour in relation to abnormal growth in fern prothalli. *Amer. J. Bot.* **42**: 245–56.

Pätau, K., 1936. Cytologische Untersuchungen an der haploid-parthenogenetischen Milbe *Pediculoides ventricosus. Zool. J.* **56**: 277–322.

Pätau, K., 1937. SAT-Chromosom und Spiralstrucktur der Chromosomen der extra-kapsulären Körper (*Merodinium* sp) von *Colozoum inerme. Cytologia Fujii Jub. Vol.* 667–80.

Pätau, K., 1948. X-segregation and heterochromasy in the spider *Aranea reaumuri. Heredity* **11**: 77–100.

Pavan, C. and Ficq, A., 1957. Autoradiography of polytene chromosomes of *Rhynchosciara angelae* at different stages of larval development. *Nature* **180**: 983.

Peacock, W. J., 1963. Chromosome duplication and structure as determined by autoradiography. *P.N.A.S.* **49**: 793–801.

Pearson, P. L., Bobrow, M. and Vosa, C. G., 1970. Technique for identifying Y chromosomes in human interphase nuclei. *Nature* **226**: 79–80.

Pelc, S. R., 1947. Autoradiograph technique. *Nature* **160**: 749–50.

Pelc, S. R., 1956. The stripping-film technique of autoradiography. *J. App. Rad & Isotopes* **1**: 172–77.

Pelc, S. R. and Howard, A., 1952. Chromosome metabolism as shown by autoradiographs. *Exp. Cell Res.* (Suppl. 2): 269–278.

Pelling, C., 1959. Chromosomal synthesis of ribonucleic acid as shown by incorporation of uridine labelled with tritium. *Nature* (London). **184**: 655–56.

Pelling, C., 1964. Ribonukleinsäure – Synthese der Reisenchromosomen. Autoradiographische Untersuchungen an *Chironomous tentans*. *Chromosoma* (Berlin) 15: 71–122.

Perry, P. and Wolff, S., 1974. A new Giemsa method for the differential staining of sister chromatids. *Nature* **251**: 156–158.

Peto, F. H., 1935. Associations of somatic chromosomes induced by heat and chloral hydrate treatments. *Canad. J. Res. C.* **13**: 301–14.

Plaut, W., Nash D. and Fanning, F., 1966. Ordered replication of DNA in polytene chromosomes of *Drosophila melanogaster. J. molec. Biol.* **16**: 85–93.

Polani, P. E., 1972. Centromere localization at meiosis and the position of chiasmata in the male and female mouse. *Chromosoma* **36**: 343–374.

Pollister, A. W., 1939. Centrioles and chromosomes in the atypical spermatogenesis of *Vivipara. P.N.A.S.* **25**: 189–95.

Poulson, D. F. and Metz, C. W., 1938. Studies on the structure of nucleolus-forming regions and related structures in the giant salivary gland chromosomes of Diptera. *J. Morph.* **63**: 363–95.

Proescher, F. and Arkush, A. S., 1928. Metallic lakes of the oxazines (gallimin blue, gallocyanin and coelestrin blue) as nuclear stain substitutes for haematoxylin. *S.T.* **3**: 28–38.

Prokofieva, Alexandra, 1935. On the chromosome morphology of certain Amphibia. *Cytologia* **6**: 148–64.

Ramanna, M. S., 1973. Euparal as a mounting medium for preserving fluorescence of aniline blue in plant material. *S.T.* **48**: 103.

Randolph, L. F., 1932. Some effects of high temperature on polyploidy and other variations in maize. *P.N.A.S.* **18**: 222–9.

Randolph, L. F., 1935. A new fixing fluid and a revised schedule for the paraffin method in plant cytology. *S.T.* **10**: 95–6.

Randolph, L. F., 1940. Card mounts for handling root tips in the paraffin method. *S.T.* **15**: 45–8.

Rasch, E. and Woodard, J. W., 1959. Basic proteins of plant nuclei during normal and pathological cell growth. *J. Biophys. Biochem. Cytol.* **6**: 263–76.

Rees, H., 1952. Asynapsis and spontaneous chromosome breakage in *Scilla. Heredity* **6**: 88–97.

Rees, H., 1955. Genotypic control of chromosome behaviour in rye. I. Inbred lines. *Heredity* **9**: Part I, 93–116.

Regan, J. D., Setlow, R. B. and Ley, R. D., 1971. Normal and defective repair of damaged DNA in human cells: a sensitive assay utilizing the photolysis of bromodeoxyuridine. *P.N.A.S.* **68**: 708–712.

Revell, S. H., 1953. Chromosome breakage by X-rays and radiomimetic substances in *Vicia. Heredity* **6**: (Suppl.) 107–24.

Richards, O. W., 1944. Phase-difference microscopy for living unstained protoplasm. *Anat. Rec.* **89**: 548.

Rigler, R., 1966. Microfluorometric characterization of intracellular nucleic acids and nucleoproteins by acridine orange. *Acta phys. Scand.* **67** (Suppl. 267): 1–122.

Riley, R. and Chapman, V., 1957. Haploids and polyhaploids in *Aegilops* and *Triticum. Heredity* **11**: 195–207.

Robinow, C., 1941. A study of the nuclear apparatus of bacteria. *P.R.S., B.* **130**: 299–324.

Rommelaere, J., Susskind, M. and Errera, M., 1973. Chromosome and chromatid exchanges in Chinese hamster cells. *Chromosoma* **41**: 243–257.

Rückert, J., 1892. Zur Entwicklungsgeschichte des Ovarialeies bei Selachiern. *Anat. Anz.* **7**: 107–58.

Rutishauser, A., 1954. Die Entwicklungserregung des Endosperms bei pseudogamen Ranunculusarten. *Mitt. naturf. Ges. Schaffhausen.* **25**: 1–45.

Rutishauser, A., 1955*a*. Das Verhalten der Chromosomes in arteigener und artfremder Umgebung. *Viertelz. Nat. Ges. Zurich.* c(1955). 17–26.

Rutishauser, A., 1955*b*. Genetics of endosperm. *Nature* **176**: 210.

Rutishauser, A., 1956. Cytogenetik des Endosperms. *Ber. schweiz. bot. Ges.* **66**: 318–35.

Rutishauser, A., 1956. Genetics of fragment chromosomes in *Trillium grandiflorum*. *Heredity* **10**: 195–204.

Rutishauser, A., 1956. Chromosome distribution and spontaneous chromosome breakage in *Trillium grandiflorum*. *Heredity* **10**: 367–407.

Rutishauser, A. and Hunziker, H. R., 1950. Untersuchungen über die cytologie des endosperms. *Arch. Klaus-Stift. Vererb Forsch.* **25**: 477–83.

Rutishauser, A. and La Cour, L. F., 1956. Spontaneous chromosome breakage in hybrid endosperms. *Chromosoma* **8**: 317–40.

Rutishauser, A. and La Cour, L. F., 1956. Spontaneous chromosome breakage in endosperm. *Nature* **177**: 324–25.

Sachs, L., 1952. Polyploid evolution and mammalian chromosomes. *Heredity* **6**: 357–64.

Sachs, L., 1953. The giant sex chromosomes in the mammal *Microtus agrestis*. *Heredity* **7**: 227–38.

Sachs, L. and Gallily, R., 1955. The chromosomes and transplantability of tumours. I. Fundamental chromosome numbers and strain specificity in ascites tumours. *J. Nat. Cancer Inst.* **15**: 1267–89.

Sachs, L. and Gallily, R., 1956. The chromosomes and transplantability of tumours. III. *J. Nat. Cancer Inst.* **16**: 1083–93.

Saez, F. A., 1952. Differentiation of meiotic heterochromatin by microspectrophotometric techniques. *Anat. Rec.* **113**: 571.

Sakamura, T., 1927. Fixierung von Chromosomen mit siedendem Wasser. *Bot. Mag. Tokyo* **41**: 59–64.

Sasaki, M. and Makino, S., 1965. The meiotic chromosomes of man. *Chromosoma* **16**: 637–651.

Sato, D., 1937. Polymorphism of karyotypes in *Galanthus* with special reference to the SAT-chromosome. *Bot. Mag. Tokyo* **51**: 242–50.

Sauerland, H., 1956. Quantitative Untersuchungen von Röntgeneffekten nach Bestrahlung Verschiedener Meiosisstadien bei *Lilium candidum* L. *Chromosoma* **7**: 627–54.

Savage, J. R. K., 1957. Celloidin membranes in pollen tube technique. *S.T.* **32**: 283–85.

Savage, J. R. K., 1966. Double staining for comparative measurements, in squash preparations. *S.T.* **42**: 19–21.

Sax, Karl, 1938. Chromosome aberrations induced by X-rays. *Genetics* **23**: 494–516.

Sax, Karl, 1940. An analysis of X-ray induced chromosomal aberrations in *Tradescantia*. *Genetics* **25**: 41–68.

Sax, K. and Beal, J. M., 1934. Chromosomes of the Cycadales. *J. Arnold Arbor.* **15**: 255–8.

Sax, K. and Enzmann, E. V., 1939. The effect of temperature on X-ray induced chromosome aberrations. *P.N.A.S.* **25**: 397–405.

Sax, K. and Humphrey, L. M., 1934. Structure of meiotic chromosomes in microsporogenesis of *Tradescantia. Bot. Gaz.* **96**: 353–61.

Sax, K. and King, E. D., 1955. An X-ray analysis of chromosome duplication. *P.N.A.S.* **41**: 150–55.

Sax, K., King, E. D. and Luippold, H., 1955. The effect of fractionated X-ray dosage on the frequency of chromatid and chromosome aberrations. *Radiation Res.* **2**: 171–79.

Sax, K. and Luippold, H., 1952. The effect of fractional X-ray dosage on the frequency of chromosome aberrations. *Heredity* **6**: 127–31; 131–32.

Sax, K. and Sax, H. J., 1933. Chromosome number and morphology in the conifers. *J. Arnold Arbor.* **14**: 356–75.

Schaede, R., 1930. Über die Struktur des ruhenden Kernes. *Ber. d. bot. Ges.* **48**: 342–8.

Schmuck, M. L. and Metz, C. W., 1928. A method for the study of chromosomes in entire insect eggs. *Science* **74**: 600–1.

Schnedl, W., 1971. Analysis of the human karyotype using a reassociation technique. *Chromosoma* **34**: 448–454.

Schneider, W. C., 1945. Phosphorus compounds in animal tissues. Extraction and estimation of desoxypentose nucleic acid and of pentose nucleic acid. *J. biol. Chem.* **161**: 293–303.

Schrader, F., 1960. Cytology and evolutionary implications of aberrant chromosome behaviour in the harlequin lobe of some Pentatomidae (Heteroptera). *Chromosoma* **11**: 103–128.

Schreiner, A. and K. E., 1916a. Neue Studien über die Chromatinreifung der Geschlechtszellen. 1. *Arch. Bio. Paris* **22**: 1–69.

Schweizer, D., 1973. Differential staining of plant chromosomes with Giemsa. *Chromosoma* **40**: 307–320.

Scott, A. C., 1936. Haploidy and aberrant spermatogenesis in a coleopteran, *Micromalthus debilis. J. Morph.* **59**: 485–516.

Seabright, M., 1972. The use of proteolytic enzymes for the mapping of structural rearrangements in the chromosomes of man. *Chromosoma* **36**: 204–210.

Sears, E. R., 1937. Cytological phenomena concerned with self-sterility in flowering plants. *Genetics* **22**: 130–81.

Semmens, C. J. and Bhaduri, P. N., 1941. Staining the nucleolus. *S.T.* **16**: 119–20.

Serra, J. A., 1946. Histochemical tests for proteins and amino acids. *S.T.* **21**: 5–18.

Serra, J. A., 1958. A method for the cytochemical detection of masked lipids. *Rev. Port. Zool. Biol. Gen.* **1**: 109–29.

Sharma, A. K. and Bal, A. K., 1953. Coumarin in chromosome analysis. *S.T.* **28**: 255–57.

Sharman, G. B. and Barber, H. N., 1952. Multiple sex-chromosomes in the Marsupial, *Potorous. Heredity* **6**: 345–55.

Shaver, E. L. and Carr, D. H., 1967. Chromosome abnormalities in rabbit blastocysts following delayed fertilization. *J. Reprod. Fertil.* **14**: 415–420.

Shaw, G. W., 1958. Adhesion loci in the differentiated heterochromatin of *Trillium* species. *Chromosoma* **9**: 292–304.

Shigenaga, M., 1937b. An experimental study in the abnormal nuclear and cell divisions in living cells. *Cytologia, Fujii Jub. Vol.* 464–78.

Shinke, N., 1937a. An experimental study on the structure of living nuclei in the resting stage. *Cytologia, Fujii Jub. Vol.* 449–63.

Shinke, N., Ishida, M. R. and Ueda, K., 1957. A study of the Feulgen reaction of plant cells. *Proc. int. Genet. Symp. Cytologia* suppl. pp. 156–61.

Singleton, J. R., 1953. Chromosome morphology and the chromosome cycle in the ascus of *Neurospora crassa. Am. J. Bot.* **40**: 124–44.

Sirlin, J. L. and Knight, G. R., 1960. Chromosomal synthesis of protein. *Exp. Cell Res.* **19**: 210–219.

Skovsted, A., 1939. Cytological studies in twin plants. *C.R. Lab. Carlsb.* **22**: 427–45.

Slifer, E. H., 1945. Removing the shell from living grasshopper eggs. *Science* **102**: 282.

Smith, F. H., 1934. The use of picric acid with the Gram stain in plant cytology. *S.T.* **9**: 95–6.

Smith, H. H., Fussell, D. P. and Kugelman, B. H. Partial synchronization of nuclear divisions in root meristems with 5-aminouracil. *Science* **142**: 595–596.

Smith, L., 1946. Haploidy in einkorn. *J. agric. Res.* **73**: 291–301.

Smith, S. G., 1940. A new embedding schedule for insect cytology. *S.T.* **15**: 175–6.

Smith, S. G., 1941. A new form of spruce sawfly identified by means of its cytology and parthenogenesis. *Sci. Agric.* **21**: 244–305.

Smith, S. G., 1943. Techniques for the study of insect chromosomes. *Canad. Entome.* **75**: 21–34.

Smith, S. G., 1952. The cytology of some tenebrionid beetles (Coleoptera). *J. Morph.* **91**: 325–64.

Smith, S. G., 1966. Natural hybridization in the Coccinellid genus *Chilocorus. Chromosoma* **18**: 380–406.

Snell, G. D., 1939. The induction by irradiation with neutrons of hereditary changes in mice. *P.N.A.S.* **25**: 11–14.

Snoad, B., 1955. The action of infra-red upon chiasma formation. *Chromosoma* **7**: 451–59.

Snow, R., 1963. Alcoholic hydrochloric acid–carmine as a stain for chromosomes in squash preparations. *S.T.* **38**: 9–13.

Sonnenblick, B. P., 1940. Cytology and development of the embryos of X-rayed adult *Drosophila melanogaster. P.N.A.S.* **26**: 373–81.

Sparrow, A. H. and Christensen, E., 1953. Tolerance of certain plants to chronic exposure to gamma radiation from cobalt-60. *Science* **118**: 697–98.

Sparrow, A. H., Moses, M. J. and Dubow, R. J., 1952. Relationship between ionizing radiation, chromosome breakage and certain other nuclear disturbances. *Exp. Cell Res.* (Suppl.) **2**: 245–67.

Sparrow, A. H. and Singleton, W. R., 1953. The use of radiocobalt as a source of gamma rays and some effects of chronic irradiation on growing plants. *Amer. Nat.* **87**: 29–48.

Spearing, J. K., 1937. Cytological studies of the Myxophyceae. *Arch. Protistenk.* **89**: 209–78.

Stacey, M., Deriaz, R. E., Teece, E. G. and Wiggins, L. F., 1946. Chemistry of the Feulgen and Dische nucleal reactions. *Nature* **157**: 740–1.

Stadler, L. J., 1928. Some genetic effects of X-rays in plants. *J. Hered.* **21**: 3–19.

Stadler, L. J., 1931. The experimental modification of heredity in crop plants. I. Induced chromosomal irregularities. *Sci. Agric.* **11**: 557–72.

Staiger, H., 1950. Cytologische und Morphologische Untersuchungen zur Determination der Nähreier bei Prosobranchiern. *Z. Zellforsch.* **5**: 496–549.

Stebbins, G. L. and Ellerton, S., 1939. Structural hybridity in *Paeonia cali-fornica*, etc. *J. Genet.* **38**: 1–36.

Stebbins, G. L. and Jenkins, J. A., 1939. Aposporic development in the North American species of *Crepis. Genetica* **21**: 191–224.

Steffenson, D., 1953. Induction of chromosome breakage at meiosis by a magnesium deficiency in *Tradescantia. P.N.A.S.* **39**: 613–20.

Steffenson, D., 1955. Breakage of chromosomes in *Tradescantia* with a calcium deficiency. *P.N.A.S.* **41**: 155–60.

Stowell, R. E., 1945. Feulgen reaction for thymonucleic acid. *S.T.* **20**: 45.

Strangeways, T. S. P. and Canti, R. G., 1927. The living cell *in vitro* as shown by dark ground illumination and the changes induced in such cells by fixing reagents. *Q.J.M.S.* **71**: 1–14.

Straub, J., 1936. Untersuchungen zur Physiologie der Meiosis. II. *Zeits F. Bot.* **30**: 1–57.

Sumner, A. T., 1972. A simple technique for demonstrating centromeric hetero-chromatin. *Exp. Cell Res.* **75**: 304–306.

Sumner, A. T., Evans, H. J. and Buckland, R. A., 1971. A new technique for distinguishing between human chromosomes. *Nature New Biol.* **232**: 31–32.

Sumner, A. T., Evans, H. J. and Buckland, R. A., 1973. Mechanisms involved in the banding of chromosomes with quinacrine and Giemsa. I. The effects of fixation in methanol–acetic acid. *Exp. Cell Res.* **81**: 214–222.

Sumner, A. T. and Evans, H. J., 1973. Mechanisms involved in the banding of chromosomes with quinacrine and Giemsa. II. The interaction of the dyes with the chromosomal components. *Exp. Cell Res.* **81**: 223–236.

Sunderland, N. and McLeish, J., 1961. Nucleic acid content and concentration in root cells of higher plants. *Exp. Cell Res.* **24**: 541–54.

Sunderland, N. and Wicks, F. M., 1971. Embryoid formation in pollen grains of *Nicotiana tabacum. J. Exp. Bot.* **22**: 213–226.

Suomalainen, E., 1940. Polyploidy in parthenogenetic Curculionidae. *Hereditas* **26**: 51–64.

Suomalainen, E., 1954. Zur Zytologie der parthenogenetischen Curculioniden der Schweiz. *Chromosoma* **6**: 627–655.

Suomalainen, E., 1965. On the chromosomes of the geometrid moth genus *Cidaria. Chromosoma* **16**: 166–184.

Suomalainen, E., 1966. Achiasmatische Oogenese bei Trichopteren. *Chromosoma* **18**: 201–207.

Suomalainen, H. O. T., 1952. Localization of chiasmata in the light of observa-tions on the spermatogenesis of certain *Neuroptera. Ann. Zool. Soc. Zool. Botan. Fennicae 'Venamo'* **15**: 1–104.

Svardson, G., 1945. Chromosome studies on Salmonidae. 23 *Rep. Swedish State Inst., Drottningholm.*

Swanson, C. P., 1940. The use of acenaphthene in pollen-tube technique. *S.T.* **15**: 49–52.

Swanson, C. P., 1942. The effects of ultraviolet and X-ray treatment on the pollen-tube chromosomes of *Tradescantia. Genetics* **27**: 491–503.

Swanson, C. P., 1943. Differential sensitivity of prophase pollen tube chromo-somes to X-rays and ultraviolet radiation. *J. gen. Physiol.* **26**: 485–94.

Swanson, C. P., 1943. The behaviour of meiotic prophase chromosomes as re-vealed through the use of high temperature. *Am. J. Bot.* **30**: 422–28.

Swanson, C. P., 1944. X-ray and ultraviolet studies on pollen-tube chromo-somes. *Genetics* **29**: 61–68.

Swanson, C. P., 1949. Further studies on the effect of infrared on X-ray induced aberrations in *Tradescantia. P.N.A.S.* **35**: 237–44.

Swanson, C. P., 1955. Relative effects of qualitatively different ionizing radiations on the production of chromatid aberrations in air and nitrogen. *Genetics* **40**: 193–203.

Swanson, C. P. and Schwartz, D., 1953. Effect of X-rays on chromatid aberrations in air and in nitrogen. *P.N.A.S.* **39**: 1241–50.

Swim, H. E. and Parker, R. F., 1957. Properties acquired by cells maintained in continuous culture. *N.Y. Ac. Sci. Spec. Pub.* **5**: 351–55. Cell biology, N.A. and viruses.

Taft, E. B., 1951. The problem of a standardized technique for methyl-green-pyronin stain. *S.T.* **26**: 205–12.

Tanaka, M., 1937. A method of rendering difficult interspecific crossing successful by means of X-rays. *Bot. and Zool.* **5**: 1567.

Tarkowski, A. K., 1966. An air drying method for chromosome preparations from mouse eggs. *Cytogenetics* **5**: 394–400.

Tartar, V. and Chen, T-T. 1941. Mating reactions of enucleate fragments in *Paramecium bursaria. Biol. Bull.* **80**: 130–8.

Taylor, J. H., 1958. Sister chromatid exchanges in tritium-labelled chromosomes. *Genetics* **43**: 515–529.

Taylor, J. H., 1962. Chromosome reproduction. *Int. Rev. Cytol.* **13**: 39–73.

Taylor, J. H., 1974. Units of DNA replication in chromosomes of *eukaryotes. Int. Rev. Cytol.* **37**: 1–20.

Taylor, J. H. and McMaster, R. D., 1954. Autoradiographic and microphotometric studies of deoxyribonucleic acid during microgametogenesis in *Lilium longiflorun. Chromosoma* **6**: 489–521.

Taylor, J. H., Woods, P. S. and Hughes, W. L., 1957. The organization and duplication of chromosomes as revealed by autoradiographic studies using tritium-labelled thymidine. *P.N.A.S.* **43**: 122–28.

Taylor, W. R., 1924. The smear method for plant cytology. *Bot. Gaz.* **78**: 236–8.

Therman, E., 1951. The effect of indole-3-acetic acid on resting plant nuclei. I. *Allium cepa. Ann. Acad. Sci. Fen.* (A) IV. **16**: 1–40.

Thoday, J. M., 1942. The effects of ionizing radiations on the chromosomes of *Tradescantia bracteata.* A comparison between neutrons and X-rays. *J. Genet.* **43**: 189–210.

Thoday, J. M., 1953. Sister-union isolocus breaks in irradiated *Vicia faba* etc. *Heredity* **6**: (Suppl.).

Thoday, J. M. and Read, J., 1947. Effect of oxygen on the frequency of chromosome aberrations produced by X-rays. *Nature* **160**: 608.

Thoday, J. M. and Read, J., 1949. Effect of oxygen on the frequency of chromosome aberrations produced by alpha rays. *Nature* **163**: 133–34.

Thomas, P. T., 1940. The aceto-carmine method for fruit material. *S.T.* **15**: 167–72.

Thomas, P. T., 1945. Supercharging the nucleus. *Discovery* **6**: 376–382.

Thomas, P. T. and Revell, S. H., 1946. Secondary association and heterochromatic attraction. I. *Cicer. Ann. Bot.* **10**: 159–64.

Tjio, J. H. and Levan, A., 1950. The use of oxyquinoline in chromosome analysis. *An. Estac. Exp. Aula. Dei.* **2**: 21–64.

Tjio, J. H. and Levan, A., 1954. Some experiences with acetic orcein. *An Estac. Exp. Aula Dei.* **3**: 225–28.

Tjio, J. H. and Levan, A., 1956. The chromosome number of man. *Hereditas* **42**: 1–6.

Tjio, J. H. and Levan, A., 1956. Comparative idiogram analysis of the rat and the Yoshida rat sarcoma. *Hereditas* **42**: 218–34.

Tjio, J. H. and Whang, J., 1962. Chromosome preparation of bone marrow cells. Without prior *in vitro* culture or *in vivo* colchicine administration. *S.T.* **37**: 17–20.

Trosko, J. E. and Wolfe, S., 1965. Strandedness of *Vicia faba* chromosomes as revealed by enzyme digestion studies. *J. Cell Biol.* **26**: 125–135.

Tschermak-Woess, E., 1956. Karyologische Pflanzenanatomie. *Protoplasma* **46**: 799–834.

Tschermak-Woess, E. and Dolezal, R., 1952. Durch Seitenwurzelbildung induzierte und spontane Mitosen in den Dauergeweben der Wurzel. *Osterreichischen Botanischen Zeitschrift* **100**: 3, 358–402.

Tuan, H., 1930. Picric acid as a destaining agent for iron alum. *S.T.* **5**: 135–8.

Ulrich, H., 1957. Die Strahlenempfindlichkeit von Zellkern und Plasma u.s.w. *Zool. Anz.* **19** (Suppl.).

Upcott, M. B., 1936a. The origin and behaviour of chiasmata, XII. *Eremurus spectabilis*. *Cytologia* **7**: 118–30.

Upcott, M. B., 1936b. The mechanics of mitosis in the pollen-tube of *Tulipa*. *P.R.S., B.* **121**: 207–20.

Upcott, M. B. and La Cour, L., 1936. The genetic structure of *Tulipa*. I. A chromosome survey. *J. Genet.* **33**: 353–72.

Upcott, M. B., 1939. The genetic structure of *Tulipa*. III. Meiosis in polyploids. *J. Genet.* **37**: 303–39.

Utakoji, T., 1972. Differential staining patterns of human chromosomes treated with potassium permanganate. *Nature* **239**: 168–170.

Vaarama, A., 1949. Meiosis in moss species of the family Grimmiaceae. *Port. Acta Biol.* Series *A*. R. B. Goldschmidt vol. pp. 47–48.

Vaarama, A., 1950. Studies on chromosome numbers and certain meiotic features of several Finnish moss species. *Bot. Not.* pp. 239–59.

Vaarama, A., 1954. Cytological observation on *Pleurozium schreberi*, with special reference to centromere evolution. *Ann. Bot. Soc. 'Vanamo'* **28**: 1–59.

Vosa, C. G., 1961. A modified aceto-orcein method for pollen mother cells. *Caryologia* **14**: 107–10.

Vosa, C. G., 1970. Heterochromatin recognition with fluorochromes. *Chromosoma* **30**: 366–372.

Vosa, C. G., 1973. Heterochromatin recognition and analysis of chromosome variation in *Scilla sibirica*. *Chromosoma* **43**: 269–278.

Vosa, C. G. and Marchi, P., 1972. Quinacrine fluorescence and Giemsa staining in plants. *Nature (Lond.) New Biol.* **237**: 191–192.

Wada, B., 1935. Mikrurgische Untersuchungen lebender Zellen in der Teilung II. *Cytologia* **6**: 381–406.

Waddington, C. H. and Kriebel, T., 1935. A 'dope' for embedding wax. *Nature* **136**: 685.

Wahrman, J. and Goitein, R., 1972. Hybridization in nature between two chromosome forms of spiny mice. In: *Chromosomes Today*. pp. 228–237.

Walters, J. L., 1956. Spontaneous meiotic chromosome breakage in many natural populations of *Paeonia californica*. *Am. J. Bot.* **43**: 342–54.

Walters, M. S., 1954. A study of pseudobivalents in meiosis of two interspecific hybrids of *Bromus*. *Amer. J. Bot.* **41**: 160–71.

Walters, M. S., 1957. Studies of spontaneous chromosome breakage in interspecific hybrids of *Bromus*. *Univ. Calif. Publ. Bot.* **28**: 335–447.

Wang, H. C. and Fedoroff, S., 1972. Banding in human chromosomes treated with trypsin. *Nature New Biol.* **235**: 52–53.

Waring, M. and Britten, R. J., 1966. Nucleotide sequence repetition: a rapidly reassociating fraction of mouse DNA. *Science* **154**: 791 et seq.

Warmke, H. E., 1935. A permanent root-tip smear method. *S.T.* **10**: 101–3.

Warmke, H. E., 1941. A section-smear method for plant cytology. *S.T.* **16**: 9–12.

Warmke, H. E., 1946. Pre-cooling combined with chrom-osmo-acetic fixation, etc. *S.T.* **21**: 87–9.

Waterman, H. C., 1939. The preparation of hardened embedding paraffins having low melting points. *S.T.* **14**: 55–62.

Watkins, A. E., 1925. Genetical and cytological studies in wheat, II. *J. Genet.* **15**: 323–66.

Webber, J. M., 1940. Polyembryony. *Bot. Rev.* **6**: 575–98.

Weisblum, B. and de Haseth, P., 1972. Quinacrine, a chromosome stain specific for deoxyadenylate–deoxythymidylate-rich regions in DNA. *P.N.A.S.* **69**: 629–632.

Wells, B. and La Cour, L. F., 1971. A technique for studying one and the same section of a cell in sequence with the light and electron microscope. *J. Microsc.* **93**: 43–48.

Westergaard, M., 1940. Studies on cytology and sex determination in polyploid forms of *Melandrium album. Dansk. Bot. Ark.* 10 (5).

Whitaker, W., 1939. The use of the Feulgen technique with certain plant materials. *S.T.* **14**: 13–16.

White, M. J. D., 1935. The effects of X-rays on mitosis in the spermatogonial divisions of *Locusta migratoria* L. *P.R.S., B.* **119**: 61–84.

White, M. J. D., 1936. The chromosome cycle of *Ascaris megalocephala. Nature* **137**: 783.

White, M. J. D., 1938. A new and anomalous type of meiosis in a mantid, *Callimantis antillarum* Saussure. *P.R.S., B.* **125**: 516–23.

White, M. J. D., 1940. The heteropycnosis of sex chromosomes and its interpretation in terms of spiral structure. *J. Genet.* **40**: 67–82.

White, M. J. D., 1946. The cytology of the Cecidomyidae. *J. Morphol.* **78**: 201–19.

White, M. J. D., 1965. Principles of karyotype evolution in animals. In: *Genetics Today* **2**: 391–7. Geerts, S. J. (ed.). Oxford (Pergamon Press).

Wickbom, T., 1945. Cytological studies on Dipnoi, Urodela, Anura, and *Emys. Hereditas* **31**: 241–346.

Wilkins, M. H. F., 1956. Physical studies of the molecular structure of deoxyribose nucleic acid and nucleoprotein. *Col Spr. Harb. Symp. Quant. Biol.* **21**: 75–90.

Williams, W. and Dowrick, G. J., 1958. The uptake and distribution of radioactive phosphorus (P[32]) in relation to the mutation rate in plants. *J. hort. Sci.* **33**: 80–95.

Wilson, I. M., 1937. A contribution to the study of the nuclei of *Peziza tutilans* Fries. *Ann. Bot.* **1**: 655–72.

Wischnitzer, S., 1957. A study of the lateral loop chromosomes of amphibian oöcytes by phase contrast microscopy. *Am. J. Anat.* **101**: 135–57.

Witschi, E., 1933. Contributions to the cytology of amphibian germ cells. I. *Cytologia* **4**: 174–81.

Wolf, B. E., 1957. Temperaturabhängige Allozyklie des Polytänen X-Chromosoms in den Kernen der Somazellen von *Phryne cincta. Chromosoma* **8**: 396–435.

Wolff, S., 1953. Some aspects of the chemical protection against radiation damage to *Vicia faba* chromosomes. *Genetics* **39**: 356–64.

Woodard, J., Rasch, E. and Swift, H., 1961. Nucleic acid and protein metabolism during the mitotic cycle in *Vicia faba. J. Biophys. Biochem. Cytol.* **9**: 445–462.

Woodard, J. and Swift, H., 1964. The DNA content of cold-treated chromosomes. *Exp. Cell Res.* **34**: 131–137.

Wroblewzka, J. and Dyban, A. P., 1969. Chromosome preparations from mouse embryos during early organogenesis: dissociation after fixation, followed by air drying. *S.T.* **44**: 147–150.

Wylie, A. P., 1952. The history of the garden narcissi. *Heredity* **6**: 137–56.

Wylie, A. P., 1957. Chromosome numbers of mosses. *Trans. Brit. Bryol. Soc.* **3**: 260–78.

Yamasha, G. and Nomura, K., 1939. Über den Einfluss der Wasserstoffionenkonzentration der Neutralsalze auf die Vitalfärbung. u.s.w. *Sc. Rep. Tokyo Univ. Lit. Sci.* **4**: 27–42.

Yasuda, E., 1934. Self-incompatibility in *Petunia violocea. Bull. Imp. Coll. Ag. Morika* **20**: 1–95.

Yasui, K., 1933. Ethyl alcohol as a fixative for smear materials. *Cytologia* **5**: 140–5.

Young, J. Z., 1935. Osmotic pressure of fixing solutions. *Nature* **135**: 823.

Yost, H. T., 1951. The frequency of X-ray induced chromosome aberrations in *Tradescantia* as modified by near infrared radiation. *Genetics* **36**: 176–84.

Yunis, J. J. and Yasmineh, W. G., 1971. Heterochromatin, satellite DNA and cell function. *Science* **174**: 1200–1210.

Zakharov, A. F. and Egolina, N. A., 1972. Differential spiralization along mammalian mitotic chromosomes. I. BUdR revealed differentiation in Chinese hamster chromosomes. *Chromosoma* **38**: 341–365.

Zakharov, A. F., Baranovskaya, L. J., Abraimov, A. J., Benjusck, V. A., Deminsteva, V. S. and Oblapenko, N. G., 1974. Differential spiralization along mammalian mitotic chromosomes. II. 5-Bromodeoxyuridine and 5-bromodeoxycytidine revealed differentiation in human chromosomes. *Chromosoma* **44**: 343–359.

Zirkle, C., 1930. Use of N-butyl alcohol in dehydrating woody tissues for paraffin embedding. *Science* **71**: 103–4.

Zirkle, C., 1940. Combined fixing, staining and mounting media. *S.T.* **15**: 139–53.

INDEX

Abraimov 79
Acarina 103
Acenaphthene 176
Acetic acid 31 *sqq.*, 51, 89, 111, 113, 114, 118, 138
Acetic alcohol 37, 58, 61, 84, 92, 114, 124 *sqq.*
Acetic stains, *see* Stain-fixatives
Aceto-carmine, *see* Carmine
Achromatic condenser 23
Acid haematein 53
Acomys 105
Acridine orange 64, 140
Acridines 72
Acrocentric 146
Actinomycin 52
Adenosine-^{3}H 97
Adkisson 133
Agammaglobulin 137
Agapanthus 66
Agar 29, 82, 83, 119
Agropyron 106
Albumen 43, 92, 119, 123
Alcohol:
  butyl, n 42, 122, 129–30; ethyl (ethanol) 33, 35, 41, 44, 48, 84, 111, 114, 133; iso-propyl 42; methyl 53, 61, 114, 131, 137; tertiary butyl 42
Alfert 47, 52, 95
Algae 37, 136, 137
Alkylating agents 73
Allen 30, 34
Allium 72–5, 107, 108
Allolobophora 55
Alpha-rays 70, 76
Alum 33, 37, 47, 118
Amaryllis 82
Ambystoma 50
5-Aminouracil 75
Ammonium hydroxide (NH₄OH) 50, 51
Ammonium molybdate 129
Ammonium Oxalate 38
Amphibia 50, 54, 58, 104, 110
Amphipoda 103
Anacridium 103
Anaphase 78, 150; regression 78
Anderson: E. 106, 107; N. G. 54; R. L. 86, 104
Andersson 106

Angiospermae 106
Ångstrom 145
Anilin blue 53, 116, **129**
Anilin oil 127
Anilocra 103
Animal: breeding 20, 81; eggs 32; tissue 35
Annelida 55, 102
Anoxia 68
Anthers 34, 368
Antirrhinum 74
Ants 65
Aphis 104
Aplanatic condenser 25
Apochromatic objectives 22, 24
Apomixis 146
Apterona 104
Arachnida 103, 135
Aranea 103
Arc-lamp 48
Arginine 52
Arginine: test for 52, 130; reaction mixture 116
Arkush 47
'Arrhenotoky' 100
Arrighi 54, 62
Articulata 103
Ascaris 102
Ascomycetes 38, 106, 137
Aspidoproctus 104
Asynapsis 146
Atebrin 133
Aucuba 109
Auerbach 76
Aulacantha 102
Aurantia 74
Autoradiography 53, 77, 91, *sqq.*; artefacts in 94; exposure in 93; film-stripping technique 92, 134; observation in 95; removing an autoradiograph from the slide 95
Autosomes 123, 146
Avanzi 53, 75
Avelino 62
Aves 105
Avery 74
Azure A 47
Azure B 53, 130

Babcock 107
Bacteria 47, 58, 129, 168
Baer 85
Baird 42
Bajer 28, 29, 30
Baker 54, 122
Bal 75
Balbiani 56
Baldwin 40
Balsam 48, 60
Baranetzky 50
Baranovskaya 79
Barber 28, 29, 73, 74, 80, 81, 106, 107, 110
Barer 30
Barigozzi 51
Barium hydroxide 62, 132
Basic dyes 31
Bateman 70
Battaglia 125
Bauer 28, 45, 46, 56, 57, 103
Beadle 81
Beal 106
Beams and King 81
Becker 28
Beermann 103, 104
Belanger 91
Belar 28, 29, 47, 102
Bellevalia 96
Belling 25, 36, 47
Benazzi 102
Benda 32, 144
Benjusch 79
Berenbaum 53
Bernal 98
Bernardo 40
Beta-rays 68, 70
Bhaduri 47, 165
Bianchi 95
Bibio 57
Bird 73
Birefringence 30
Bishop 69, 83
Bismarck Brown 38, 55
Blakeslee, 74
Bleaching 44, 46, 49, 115, 119, 121
Bloch 52
Blood 53, 59, 114, 136; culture method 136; diseases 59; fixative 114; peripheral 60; plasma 176; worm 57
Bobrow 61
Boche 29
Boiling points 111, 112
Bolomey 70
Bone marrow 137
Boothroyd 79
Bose 115
Bosemark 106

Bougin 85
Bouin 33, 34; Allen's 114
Bovine serum 137
Brachet 52, 53, 130
Bradbury 42
Brazilin 47
Breckon 47, 60
Breeding experiments 20, 80, 81
Brewbaker 82
Brewen 77, 78
Brian 103
Bridges 38
Brieger 103
Bristol board 27, 144
Britten 61
Bromelin 133
Bromide paper 89
5-Bromodeoxycytidine (BCdR) 79
5-Bromodeoxyuridine (BUdR) 64, 77, 78, 79
a-Bromonaphthaline 26, 39, 75
Bromus 73, 106
Brown: Robert 19; S. W. 66
Brownian movement 59
Bryophyta 106
Buccal 61
Buchholz 84
Buck 29
Buckland 62
Bulk: fixation 31 *sqq.*; staining 45
Burch 30
Butcher 60

Calcium chloride 83
Calcium nitrate 82
Callan 28, 51, 54, 58, 79, 80, 96, 97, 103, 104, 110
Camargo 52
Camera 87; lucida 22, 27
Campanula 85, 108
Canada balsam, *see* Balsam
Cancer 60, 68, 105
Cane sugar, *see* Sucrose
Canti 29
Capinpin 47
Carbol fuchsin 116, 133
Carbon, decolorising 46, 115
Carlson 69, 118
Carmine 26, 36, 37, 50, 51, 58, 59, 115, 123, 134
Carnoy 33, 39, 114, 133
Carnoy-lebrun 114
Carr 60, 133
Carson 104
Caspersson 29, 30, 51, 52, 54, 133
Catcheside 69, 70
Cedarwood oil 26, 38, 124
Celarier 84

Celestin blue B 47
Cell 19, 146; anaesthesia 78; death 48; fixation 31, 36 *sqq.*; histochemistry of, 51 *sqq.*; separation 39
Celloidin 92, 131
Cellophane 83, 136
Cellulose 32, 35
Cell-walls 33, 39
Centrifuging 54, 81, 137 *sqq.*
Centromere 61, 62, 66, 77, 103, 104, 146
Centromeric banding (C-banding) 61, 131
Centrosome 146
Cereals 34, 85
Chameleon 105
Chamot 25
Champy 33, 113
Chapelle 64
Chara 19, 28, 108
Characeae 105
Chayen 39, 53
Chelating agents 77
Chemical formulae 111, 112
Chemistry 20, 33, 51, 52
Chen 86, 102
Chiasma 146, 147
Chilocorus 104
Chimaera 147
Chinese hamster 139
Chironomus 28, 29, 57, 96, 104, 125
Clorazol black E 38
Chloroform 41, 47, 110, 114, 121, 125, 144
Chlorophyseae 105
Chondriosomes 32
Chordata 104
Chorthippus 96, 110
Chortophaga 69
Christensen 68
Chromatid 147; sister 63, 64; sister exchange 63; -sub 72
Chromatin 53, 63, 147; sex 131
Chromatin fibrils 79
Chrome alum 92, 118
Chromic acid 32 *sqq.*, 45, 46, 49, 53, 111, 113, 123, 128
Chromocentre 66, 147, 184
Chromosome 147, 148; abnormalities 81; B- 102, 147; basic number 146; breakage 29, 67, 68, 146; complement 147; deficiency 148; disjunction 148; pairing 66, 150; rings 147; set 147; sex 61, 131, 151; size 56, 57, 68; telocentric 146; *see also* Nucleic acid and Proteins
Chrysanthemum 46
Cicadales 106
Cicer 66
Cidaria 104
Ciliata 102
Cimex 104

Cine film 29
Cladophora 105
Clarite 48
Clarase 39
Classification 102 *sqq.*
Claude 54
Cleavage 55
Cleveland 102
Clone 148
Clove-oil 48, 128
Coagulation 31 *sqq.*
Colcemid 59, 131, 133, 138, 140
Colchicine 39, 40, 59, 74, 75, 78, 131, 133, 138, 139
Cold shocks 40
Cold treatment, *see* Temperature
Cole 47
Coleman 46, 51, 115
Coleoptera 104
Coles 23, 25
Collotype 90
Collozoum 102
Colombo 103
Columba 105
Comings 62
Condenser 22; immersion 25
Configuration 148
Conger 38, 72, 83, 128
Congression 148
Conidia 67
Coniferales 65, 106
Conjugation 148
Constriction 148
Contrast 89
Cooper 103
Copepoda 103
Copying 87
Corrosive sublimate 112, 114
Cotton blue 85, 117
Coumarin 72, 75
Counter-stains 47
Cover slip 33, 37, 38, 124, 142; cleaning 118; sealing 29, 38, 118; thicknesses 24
Creighton 50
Crepis 74, 107
Cricetus 105
Critical illumination 25
Crocus 109
Crossing over 80, 148
Crustacea 103
Crystal violet (CV) 26, 33, 48, 49, 51, 55, 116, 128
Culex 59, 110
Cullis 62
Culture media (*see* Ringer) 29, 82
Customs and Excise 44
Cyanophyceae 105
Cyratacanthacris 96

Cytase 39, 62
Cytidine-³H 97
Cytoplasm 31, 37–9, 47, 52, 54, 148

D'Amato 53, 72, 77
Daniels 70
Danon 58, 61
Daphnia 103
Das 95
Datura 74, 84
Davidson: D. 69, 76, 77; J. N. 52, 130
Davies 30
Day 38
Deeley 52
de Haseth 63
Dehydration 41, 121, 122
Delafield's haematoxylin 85, 117
De Lamater 47
Demerec 57, 104
Deminsteva 79
Dendrocoelum 102
Dermaptera 103
Dermen 75, 78
Desiccation 30, 32, 83
Desoxyribose-nucleic acid (DNA) 20, 45, 46, 47, 50–7, 147, 149, 151; annealing 61; centromeric 61; denaturation 61, 64; *in situ* characterisation 64; nuclease (DN-ase) 51, 54, 115, 130; removal of 54, 130; satellite 55; substitution 80; synthesis of 151
De Pex 132
Detergent 134
Developer 89; formulae 120
Dewey 77
Diakinesis 149
Diamond 142
Diaphragms 25
Dicentric rings 78
2, 4-Dichloro-x-naphthol 116
Diepoxybutane 72
Differential reactivity 64, 79
Differential segment 148
Differentiation (in staining) 45, 46, 128
Dimethyl hydantoin formaldehyde resin 47
Diminution 148
Dioxan 42, 48, 121
Diplo-chromosomes 68, 80
Diploid 148
Diplotene 58, 149
Diprion 104
Diptera 38, 56, 69, 104, 141
Dissection 23, 34, 57
Doniach 92
Doroshenko 83
Dorsey 80
Dosage, *see* X-rays, Drugs

Double helix 51
Dowrick 46, 71, 80
Drawings 27, 99, 144
Drew, 105
Drosophila 20, 29, 56, 57, 69, 71, 96, 97, 104, 119, 125; food 119
Drugs 78, 80
Dry ice 38
Dufrenoy 42
Dugezia 102
Duryee 58
Dustin 78
Dutrillaux 62
Dyban 60
Dyer 106
Dyes, *see also* Staining, 46

Eberle 105
Ebstein 52
Echinoidea 102
Echinoderms 86
Ectocyclops 103
Edwards 105
Eggs 32, 43, 55, 86, 110
Eggshells 33, 55
Egolina 79
Ehrenberg 72
Ehrlich's haematoxylin 135
Eigsti 78
Eisenia 102, 134
Ellenhorn 30
Elliott 38, 80
Embedding 30, 41, 42, 121, 122
Embryo sac 40, 43, 53, 65, 81, 124
Emerson 84
Emsweller 39, 81
Endicott 92
Endomitosis 106, 148
Endosperm 28, 65, 69, 106; squash method 126; Tween method 133
Enzyme 39, 53, 54, 130, 132
Enzymes proteolytic 62
Eosin 42, 135
Epon resin 47
Epoxides 72, 73
Erickson 54
Ernst-Schwarzenbach 105
Errera 77
Eschscholtzia 82
Esser 85
Ester wax 42, 122
8-Ethoxycaffeine 73
Ethylene glycol monoethyl ether ('Cellosolve') 122
Ethylene oxide 72
Ethylurethane 72
Euchromatin 54, 79, 80, 149, 187
Euglena 102

'Euparal' 48, 93, 124 *sqq.*, 135, 136
Evans: E. P. 47, 60; H. J. 47, 61, 62; T. C. 92; W. L. 79
Exine (solvent for) 40
Exposure 88; of autoradiographs 93
Eyepieces 22 *sqq.*, 88; micrometer 27

Fabergé 39, 42, 71, 176
Fading 48, 49, 128
Fahmy 73
Fairchild 38, 72, 83, 128
Fankhauser 86, 104
Fast green 47
Favorsky 74
Fedoroff 62
Femoral 59
Ferric acetate 37; chloride 37; nitrate 118
Fertility 82
Fertilisation 20, 55, 82 *sqq.*; self 81
Feulgen 33, 45, 46, 48–52, 55, 58, 93, 94, 127 *sqq.*
Ficin 133
Ficq 57, 71
Film: 35 mm 87; microneg. 87
Filter: barrier 25, 63; excitor 25, 63; *see also* Screen
Fixation 31 *sqq.*, 36, 37, 123 *sqq.*; air-drying 55; in autoradiography 92, 93
Fixatives 31–7, 49, 92, 93, 110, 114; stain-36 *sqq.*; warm 34
Flagellata 102
Flax 53, 129
Flemming 32, 49, 50, 113
Flower-buds 34, 42
Fluorescence 61, 64
Fluorescence microscope 22
Fluorochromes 63
Focusing 88
Fogwill 109
Foley 60
Foot and Strobell 55
Foramnifera 102, 134
Forceps 58, 142
Ford: C. E. 39, 59, 60, 138; E. H. R. 60
Forficula 103
Formaldehyde (Formalin) 33, 35, 46, 61, 111, 114, 123, 125, 130, 131, 134
Fowler 105
Freezing, *see* Temperature
Fritillaria 54, 62, 65, 79, 80, 106, 107, 109
Fuchsin: acid 42, 85, 117; basic 45, 115, 116, 126, *sqq.*; carbol 116
Fugo 60
Fungi 106, 137
Fussell 75

Gagnieu 82
Galanthus 109

Gall 54, 61, 97
Gallagher 62, 103
Gallily 60
Gallocyanin 47
Gamete 149
Gametophyte 149
Gamma-rays 68, 70
Gammarus 103
'Gammexane' 75
Ganglia 57, 63, 81, 124, 132
Gastropoda 102
Gay 71
Geard 77
Geitler 51, 105
Gelatin 38, 82, 92, 93, 95, 115
Gelei 47, 102, 129
Gene 20, 31, 69, 73, 96, 131, 187
Generative nucleus (GN) 82, 83, 84, 149
Genetic 149
Genotype 149
Germination: pollen 83 *sqq.*, 136; seeds 74
Geschwind 52
Geyer-Duszynzka 69
Ghosh 130
Gibson 62
Giemsa 62, 168
Giemsa banding (G-banding) 14, 62, 131
Giles 68, 70, 71, 87
Glands *see* Salivary
Glazing 90
Glumes 40
Glutamine 136
Glycerine 48, 83, 117, 119, 121, 122
Godward 37, 105
Goitein 105
Goldberg 52
Goodspeed 68
Gorilla 61
Gossypium 40, 74, 85
Gottschalk 66
Gramineae 83
Grauer 103
Graupner 42
Gray 70
Grell 102
Groat 48
Ground-glass 26
Growth hormones 75, 78, 80
Guanine 53
Gulick 33, 45
Gum damar 48
Gustafsson 67, 68
Guyénot 58
Gymnospermae 106

Habrobracon 104
Haemanthus 28

Haematoxylin 33, 43, 47, 55, 116, 129;
  Delafield's 85, 117
Haga 80
Hair 106
Hale 30
Halopteris 105
Hamerton 60, 163
Hance 43, 116
Hanging drop 83, 142
Haploid, *see* Diploid
Haploidy 68, 85, 86
Haque 69, 73, 79
Hardening 33, 39, 41
Harland 74, 75, 85
Haupt's adhesive 119
Hauschka 60
Hayden 68
Heat shock 80
Heating 37
Heidenhain 116
Heitz 56, 57, 125
Helleborus 109
Henderson 96, 110
Heparin sodium 137
Herbarium pollen 84
Heredity 20
Heteroauxin 75
Heterochromatin 51–4, 62, 63, 66, 72, 79,
  80, 129, 131, 133, 149; *see also* Nucleic
  acid *and* Temperature
Heterocyprius 103
Heteroptera 104
Heteropycnosis 53, 149
Heterozygote 149
Heuzer 137
Hewitt 62, 103
Hillary 46, 47, 48, 51, 125, 162
Himes 55, 129
Histone 51
Hoaglands solution 75
Hodgman 112
Hoechst dyé 64, 140
Homo 105
Homoptera 104
Hordeum 72, 74
Hot plate 40, 48, 144
Howard 96
Hsu 54, 59, 62, 77, 131
Hughes-Schrader 104
Humidity 82, 83
Humphrey: L. M. 50; R. M. 77
Hungerford 105
Hunziker 65, 126
Huskins 75, 109
Husted 102
Hyacinthus 66, 69, 80, 107, 109
Hybrid: numerical 140; structural 149
Hybridity 57

Hybridisation 84
Hydrochloric acid (HCl) 38, 39, 45 *sqq.*
Hydrogen-ion concentration (pH) 37, 50,
  82
Hydrogen peroxide 49, 119, 121
Hydrolysis 33, 45–9, 92, 123, 125 *sqq.* 136
Hydroquinone 120
8-Hydroxyquinoline 75
Hymenoptera 86, 104
Hypotonic 51, 131, 133, 137, 138, 139
Hypotonic fluids 59

Ikeda 102
Ikushima 64, 77, 78
Illumination: critical 23, 26; dark ground
  34; incident 63
Image 23
Imines 73
Immersion oil 26, 48; *see also* Oil-immer-
  sion
Imms 59
Indole-3-acetic acid 75
Infiltration 41, 121, 122
Infra-red 71, 76
Intensifying 89; Feulgen staining 126
Interchange 149
Interference colours 139
Interphase, *see* Mitosis
Interpretation 98
Iodine 34, 116, 128
5-Iododeoxyuridine (IUdR) 75
Iris 28
Iron: acetate, 37, 58; alum 29, 37, 47
Iron-aceto-carmine, *see* Carmine
Iso-electric point (IEP), 33, 50
Isopoda 103
Iso-propyl alcohol 42
Isotonic solutions 29
Isotopes, radioactive 68, **70**, 91, 93, **96, 97**
Iwanami 84

Jacobs 103
Jacobson 130
Jagoda 92
Jain 80
Jensen 83
Johansen 42, 46
John 103, 104, 110

Kahle 55, 156
Karling 28, 105, 108
Karpechenko 74, 114
'Karyokinesis' 100
Kasten 46
Kato 62
Kaufmann 52, 53, 57, **69, 71, 77,** 130
Kawaguchi 81
Kelley 52

Keyl 54
Kezer 97
Kho 85
Kihara 34, 83, 85
Kihlman 72, 73
Kim 82
King 69, 81
Kirby-Smith 70
Klinger 61, 116, 131
Klingstedt 104
Knapp 68
Knight 97
Kniphofia 73, 108
Kohn 61
Koller 32, 57, 60, 70, 77, 105, 110, 112, 124
Kostoff 74, 81
Kotval 70
Kriebel 43
Kugelmann 75
Kunitz 130
Kurabayashi 80
Kurnick 53, 116, 147
Kuwada 28, 29, 50
Kwack 82

Lacmoid 38, 39, 48, 60, 85, 115, 117, 118, 124
Lactic acid 117, 125
Lamp, mercury 22, 63
Lampbrush (chromosome) 58, 147
Lane 69
Lanolin 74, 75
Lanthanum acetate 51
Lantz 81
Larter 107
Larvae 56–58, 110
Latt 64
Lawson 104
Lea 69, 70
Leaves 40
Le Blond 91
Leedale 102
Leishmann 62, 132
Lenses 23 *sqq.*, 87
Lepidoptera 104
Leuchtenberger 52
Leucho-basic fuchin 45, 115
Leucine-³H 97
Leucocytes 51
Leucojum 28
Levan 60, 73–5, 77, 81, 107, 108, 124
Levi 91
Lewis: D. 81, 82, 84, 85; K. R. 103, 106, 110
Lewitsky 32, 53, 107, 129, 131
Ley 77
Light green 47, 85, 117, 126
Lighting 25

Lilium 40, 65, 96, 108
Lilly 72
Lima de Faria 54, 66, 79, 115
Lin 102
Linnaeus 100
Linnert 72
Linskens 85
Linum 80
Living nuclei, *see* Living chromosomes, 28 *sqq.*
Locusta 69
Lolium 78, 80
Lorbeer 106
Lotfy 75
Lucas 75
Ludwig 61, 131
Lutkov 80
Luzula 66, 106
Lycoperisicum 66, 82
Lymphocytes 60

McClintock 37, 38, 66, 106, 108
McDonald 53, 77
Maceration 39, 125; *see also* Hardening
Macgregor 51, 58, 97
McLeish 52, 72, 130
McMaster 96
Madge 42
Magnesium sulphate (MgSO₄) 82, 115, 130
Magnification 123, 99
Maheshwari 42, 43
Maio 52
Makino 105
Maleic hydrazide 72, 73
Malheiros 106
Mallory 116
Malpighian tube 56
Maltose 34
Mammalia 105
Mammals 32, 47, 59, 60, 79, 81, 105, 131, 137, 138
Man 20, 60, 105, 132, 138
Mantidae 103
Manton 106
Marchi 62, 131, 133
Marks 39, 62, 108, 127
Marrow 59, 137
Martius yellow 85, 117
Mason 25
Mather 109
Matsuura 80
Matthey 32, 105
Maturation 149
Maude 109
Mayer 43, 119, 123
Mazia 77
'Megagametophyte' 100

Mehra 84
Meiosis 20, 58, 65, 69, 71, 73, 80, 102–6, 149; *see also* PMC, SMC
Melandrium 106
Melanoplus 96
Melting points 111, 112
Mercuric chloride (corrosive sublimate) 112, 114
Mercury arc 25, 63, 76
Merogony 150
Mesostoma 102
Metanil yellow 136
Metaphase 39, 150
Methionine-^{35}S 97
Methyl alcohol, *see* alcohol
Methyl-green pyronin 53, 116, 120, 135–6, 147
Methyl violet 47
Methylal paraffin oil 42
Methylene blue 29
Metz 28, 57
Meurman 109
Meyer 40, 74
Miastor 104
Michel 30, 87
Micro-chemical tests 51
Micro-incineration 51
Micromalthus 104
Micrometer 27
Microneg 87
Microphotography 87 *sqq.*, 100
Microphotometry 52
Microscope 22 *sqq.*; interference 30; phase contrast 30, 59
Microtome 30, 43, 59, 142, 144
Microtus 64, 105
Middle lamella 39
Miles 39
Miller 54
'Milton', *see* sodium hypochlorite
Minouchi 32, 113
Mirrors 22, 27
Mirsky 51
Mitochondria 54
Mitosis 20, 28, 51, 67 *sqq.*, 102–6, 150; pre-meiotic 51, 65; synchronisation of 75
Mitotic activity 78
Modlibowska 84
Mogrunda 104
Moldex 119
Molecular weights 111, 112
Mollusca 55, 102
Mongolism 105
Monobromobenzene 75
Monocytes 60
Monosmic 150
Moore 61
Mordanting 31, 33, 37, 38

Moriwaki 62
Mosaic 150
Mother cell 150
Mounting media 26, 38, 45 *sqq.*, 95, 118
Muldal 55, 102, 110
Muller 67, 68, 69
Müntzing 67, 108
Murex 102
Mus 105
Musci 106
Mustard gas 72, 73, 76
Mutation 21, 67, 68, 71, 77; gene 67, 76
Myrmeleo tettix 103

Nagao 109
Nakamura 28, 29, 30, 50, 84
Nanometre 154
Naphthalene-acetic acid 75
Narasimhan 136
Narbel 104
Narcissus 109
Navashin 33, 107, 114
Nebel 38, 51, 74, 84
Needham 87
Needles 37, 142; tungsten 59, 65
Nematoda 102
Neuroptera 103
Neurospora 67, 106
Neutrons 71, 78
'Nevillite' 48
Newcomer 82
Newt 57, 68; *see also* Triton
Newton 46
Nicklas 28
Nicotiana 85
Nipagin 119
Nitella 105, 108
Nitric acid 28, 50
Nitrogen mustard 72, 73
Nitrous oxide (laughing gas) 74
Nitsch 85
Nogusa 104
Nomarski 30
Nomura 29
Non-disjunction 150
Norit 46, 115
Nothoscordum 107
Nowell 60
Nucleic acid 20, 33, 45 *sqq.*, 5, 61–5, 79, 80, 96, 97, 130; *see also* Heterochromatin
Nuclear sap 189
Nucleolus 33, 46, 47, 52, 54, 61, 66, 72, 89, 97, 133, 136
Nucleus 19, 52, 54, 58, 61, 78, 82, 83, 84, 131, 133, 135, 136; membrane of 58, 61
Numerical aperture, (n.a.) 23–5

Objectives 22 *sqq.*, 30, 87
Oblapenko 79
Odonata 103
Oehlkers 77
Oenothera 84
Ogawa 103
Ohnuki 51
Oil-immersion 22, 23, 24, 26
Ojima 103
Okado 62
Oligochaeta 102, 110
Olive Oil 26, 48
Olivieri 77
O'Mara 39, 40, 75
Oöcyte 55, 58, 150
Oögonia 55
Optics 22 *sqq.*
Orange G, 53, 116, 129
Orcein 38, 39, 48, 58–60, 65, 81, 115, 118, 124, 125, 126, 136, 138
Orchidaceae 106
Organisers 66, 131
Orthoptera 28, 103, 140
Osmic acid 32 *sqq.*, 46, 47, 53, 112, 113, 123, 125
Osmotic pressure 33, 34
Östergren 28, 72, 74
Osterstock 90
Ostracoda 103
Otiorrhynachus 104
Otocryptops 103
Oura 50
Ovary 55, 58, 65
Oven 42, 144
Ovules 65
Oxalis 108
Oxidising agents 33, 112
Oxygen 29, 68, 72

Pachman 63
Paeonia 73, 108
Painter 52, 56, 130
Papaver 82
Pappenheim 116, 130
Para-dichlorobenzene 74
Paraffin: liquid 26, 29, 33, 59, 125; method 35, *sqq.*, 41, 65; wax 42 *sqq.*, 55, 59, 121, 122, 130; wax (melting points) 42
Paramecium 86, 102
Pardue 54, 97
Paris 108
Partanen 106
Parthenogenesis 85, 106, 150
Pätau 102, 103
Pavan 57
Peacock 77, 78, 96
Pearce 54

Pearson 61
Pectinase 39, 62, 92
Pediculoides 103
Pelargonium 85
Pelc 47, 91, 92, 93, 94, 96, 134
Pelling 54, 57, 96
Pencils 144
Penetration (of fixative) 34 *sqq.*
Penicillin 136
Pens 144
Pentatomidae 104
Peptone 39
Perchloric acid 54
Perilaneta 103
Perry 64, 139
Petals 28, 40
Petri dish 144
Petroleum ceresin 43; jelly 83
Petunia 85
Peziza 106
Phase: changing 30; contrast 30; retardation 30
Phaseolus 60, 75
Phenol 117, 122
Phenol red 136
Phenotype 150
Phospholipids 53, 54
Photography 22, 29, 87
'Photomicrography' 87, 100
Phyrne 104
Phyllopoda 103
Physiological saline 117
Phytohemagglutin 137
Picea 106
Picric acid 33
Pipette 18, 143
Pisces 104
Pisum 73
Platinic chloride 33
Platyhelmia 102
Plaut 52, 96
Plethodon 97
Pleuorzium 106
Plumaria 105
Poa 106
Polani 105
Polarised light 30
Pollen: grain (PG) 40, 51, 69–72, 81, 82 *sqq.*, 123, 160; herbarium 32, 84; mother cell (PMC) 36, 39, 43, 50, 65, 68, 107, 108, 123; storage 73, 83, 84; tube (PT) 29, 57, 69, 70, 71, 83, 84, 85, 97 *sqq.*, 176
Pollination 83, 84, 85
Polonium 70
Polychaeta 102
Polymitosis 150
Polynemy 151

Polyoxyethylene sorbitan monoleate (Tween 80) 134
Polyploid 151
Polyploidy 20, 56, 67, 74, 80, 85
Polysaccharides 45
Polytene 56, 96, 151
Potassium: bromide 120; chloride 137, 139; cyanide 72; dichromate 33, 112, 113, 118; ferrocyanide 95; iodide 116, 128; metabisulphite 115, 120; nitrate 82; permanganate 62
Potorous 105
Potter 54
Poulson 28, 57
Pre-: treatment 50, 51; drugs 39, 74, 75, 78; meiotic-mitosis 51, 65
Precocity 151
Primula 82, 84
Prism 23, 26, 30
Pro-Chromosomes 148
Proescher 47
Projection 90
Prokofieva 110
Pronase 133
Prophase 31, 53, 55, 68, 150
Propionic acid 37, 38, 118
Protease 133
Proteins 51, 52, 54
Protista 102
Prunus 82, 84
Pteridium 106
Pteridophyta 106
Publication 99, 100
Pupation 57
Purine derivatives 72
Pyrimidine base 73
Pyronin 53, 116, 129

Quantitative method 100
Quinacrine dihydrochloride 133
Quinacrine mustard 133

Radiolaria 102
Radium 68
Radon 76
Randolph 41, 42, 80
Ranunculus 106, 107
Rasch 52, 130
Rattus 105
Razari 137
Razor 43
Read 68, 70
Recombination 20
Rectified spirit 112
Reducing (illustrations) 99
Rees 73, 81
Refractive index 26, 48
Regan 78

Reich 52
Replication 79, 151
Reptilia 105
Resolving power 23
Resorcin blue 146, 38
Re-staining 49
Restitution 157; nucleus 151
Revell 72
Rhizoids 28
Rhodophyceae 105
Ribbon (paraffin) 43, 44, 144
Ribo-nuclease (RN-ase) 53–7, 169, 174
Ribose-nucleic acid (RNA) 20, 47, 52, 54, 57, 73, 92, 96, 129, 150
Richards 30
Richardson 108
Rigler 64
Riley 68
Ringer solution 29, 32, 57, 59, 118
Ris 51
Robinow 47
Robson 76
Rogers 92
Rommelaire 77
Röntgen (r) 67 *sqq.*, 145
Root tip (RT) 34, 39, 40, 42, 51, 67, 69 *sqq.*, 107, 124, 125, 127, 134, 140
Rosaceae 39
von Rosen 90, 77
Rossenbeck 45
Rotahella 102
Rubber solution 38
Ruthenic acid 33
Rutishauser 65, 69, 80, 106, 126
Ruttle 74, 84

Sachs 60, 105
Saez 129
Sakaguchi 52, 131
Sakamura 50
Salivary glands 21, 28, 56, 57, 73, 81, 119, 123–5
Salmo 104
Sanfelice 114
Saponin 34, 113
Sasaki 105
Satellite 151
Sato 109
Sauerland 69
Savage 46, 83
Sax 50, 65, 68, 69, 84, 106, 107
Scalpel 36, 134
Schaede 28
Schiff 45
Schildkraut 52
Schistocerca 96, 103
Schizostomum 102
Schmidt 30

Schnedl 62
Schneider 36, 54
Schrader 104
Schreiners 102
Schröder 64
Schwartz 68, 73
Schweizer 62, 131
Sciara 29, 57, 104
Scilla 65, 69, 73, 82, 107
Scorpiones 103
Scott 104
Screen 25, 26, 76, 87–9; projection 90
Seabright 62, 132
Sealing medium 150
Sears 84
Secale 66, 81, 96, 108
Sections 43, 92, 121, 122, 127, 128, 129, 130, 131
Seeds 60, 67, 68, 78, 126
Selander 64
Self-incompatible 81
Self-sterility 84
Semmens 47, 126
Serra 52, 53
Setlow 78
Sex chromosome 129, 151
Sex heterozygote 151
Sexual reproduction 151
Sharma 75
Sharman 105
Shaver 60
Shaw 72
Sherratt 52
Shillaber 87
Shinke 46
Siliconed slides 60, 139
Silver grains 87, 95
Sinclair 70
Singleton 38, 68
Sirlin 97
Sister chromatid exchange 77
Sister chromatids 63, 77, 79, 139
Sister reunion 151
Skovsted 85
Slifer 55
Smearing dish 142
Smears 36 *sqq.*, 44, 49, 51, 61, 92, 126, 128, 129, 130, 131
Smith: F. H. 49; H. H. 75; L. 85; S. G. 55, 104, 122
Snoad 71
Snow 38
Sodium: acetate 116; barbiturate 116; carbonate 120; chloride 117, 118; citrate 132, 144; trisodium citrate cyanide (NaCN) 50, 51; 2, 4-dichlorophen-oxyacetate 89; hydroxide (NaOH) 50, 131, 116; hypochlorite 55, 116; hypo-sulphite 120; nitrite (NaNO₂) 59, 126; salycylate 119; sulphite 85, 120
Solanum 82, 85
Solubilities 111–12
Somers 77
Sonnenblick 55
Sorbitol 118
Sparrow 68, 69, 76
Spearing 105
Spectrophotometry 29; micro- 52, 53
Sperm 30, 55, 68, 69, 81, 190; mother cells (SMC) 145, 151; nuclei 98, 68, 71, 83
Spermatocyte 151
Spermiogenesis 96
Spindle 30, 31, 35, 39, 78, 80, 151
Spiralisation 54, 80
Spiral structure 32, 50, 151, 152
Spirogyra 105
Spontaneous breakage 78
Spore 152
Sporophyte 152
Spreading 36, 48, 53, 65, 92, 99, 124 *sqq.*
Squashes 36 *sqq.*, 48, 153, 65, 92, 99, 125 *sqq.*
Stacey 52
Stadler 67, 68
Staehly 28
Staiger 102
Stain-fixatives 36 *sqq.*, 47, 48, 55, 58, 59, 115, 123 *sqq.*
Staining 33, 38, 45 *sqq.*, 93, 123 *sqq.*; of autoradiographs 135, 136
Staminal hair 19, 28
Staniland 99
Stark 29
Static electricity 43, 94
Stebbins 108
Steedman 42, 44, 122
Steffenson 77
Steinitz 75
Stems 78
Sterility 84
Stevens 87
Stigma 82, 84
Stock 30
Stop bath 89
Storage: of material 37; of pollen 73, 73, 84
Stowell 52
Strangeways 29
Straub 80
Streptomycin 136
Strobell 55
Stropping 43
Strychnine 86
Style 82, 84, 85, 117
'Subbed' (slides) 92
Sucrose 97, 29, 50

Sulphur dioxide (SO₂) water 47, 115, 126, 127
Sulphuric acid 32, 118
Sumner 62, 131, 132
Sunderland 52, 85
Suomalainen 103, 104
Surface tension 33, 110, 112; reducing agent 34
Susskind 77
Svärdson 104
Swanson 68, 70, 71, 80, 83
Swift 45, 52, 79
Synthetic orceins 115
Szent-Györgi 17

Tadpole 57
Taft 53
Tanaka 84
Tarkowski 60
Tartar 86
Taxus 109
Taylor 7, 91, 94, 96
Telophase 54, 123, 168, 188
Temperature 34, 42, 63, 72, 79, 80, 82, 83, 94; treatment 35, 63, 72, 79, 80
Tendrils 40
Tenebrionidae 104
Terminalisation 152
Testes 35, 39, 43, 58, 59, 81, 110, 138, 140
Tetrad 152
Tetrasomic 152
Therman 75
Thionin 61, 116, 131
Thionyl chloride 47
Thoday 68, 69, 70, 71, 72
Thomas 37, 40, 66, 78, 92, 124, 129
Thorium nitrate 51
Thymidine-³H 77, 96
Tipula 104
Tissues 31, 33, 92; shrinkage of 42
Title (publication) 100
Tityus 103
Tjio 60, 75, 105, 160
Toluene 48
Toluidine blue 47, 93, 117, 127, 135
Tomopteris 102
Tradescantia 19, 28, 50, 69, 70, 71–3, 77, 80–4, 106–8
Transplantation 21
Trematoda 102
Tribromaniline 74
Trichloroacetic acid (TCA) 54, 94
Trichonympha 102
Trichoptera 104
Trillium 40, 50, 67, 69, 73, 79, 80, 82, 106, 107, 108, 109
Triodopsis 102
Triploids 81, 85

Trisodium nitrate 138
Triticum 84, 85
Tritium 77
Triton 79, 80, 86, 110
Triturus 57, 58, 81, 96, 104
Trosko 75
Trypsin 51, 132
Tschermak-Woess 106
Tuan 129
Tube-length (microscope) 23, 24
Tulipa 51, 65, 66, 73, 108, 109
Tungsten, *see* Lamp *and* Needles
Tumours 60, 124
Turbellaria 102
Turpentine 118
Tween 133; method 133, 134
Twins 85

Uber 68
Ulrich 69
Ultra-violet 29, 48, 71, 98
Unna's stain 53, 116, 130, 135
Upcott 41, 47, 51, 66, 68, 73, 82, 108, 109
Uracil 73
Uranium salts 33
Urea 34, 114
Urechis 102
Urodela 28, 104
Utakoji 62
Uvularia 80, 107

Vaarama 38, 106
Vacuum pump 34, 75, 143
Van Tieghem 142
Vegetative nuclei 149
Vendreley 45, 52
Venipuncture 60, 137
Vicia 68, 69, 72, 73, 75, 96, 97
Vignetting 90
Visser 83
Vital staining 29
Vivisection 58
Vocables 100
Vosa 61, 62, 63, 131, 133

Wada 29
Waddington 43
Wahrman 105
Wakonig 72
Walker 133
Walters: J. L. 73; M. S. 73, 106
Wang 82
Waring 61
Waring, blender 54
Warmke 39, 40
Washing 41
Water blue (fluorochrome from) 85

Water pump 143
Waterman 42
Watkins 85
Wavelengths 23, 24, 67, 71, 76
Wax, *see* Paraffin
Waymouth 52, 130
Webb 130
Weisberger 42
Weisblum 63
Well-slide 142
Wells 47, 79, 80
Westergaard 106
Whang 60
White 69, 88, 102–4, 110
Wickbom 110
Wicks 85
Wilkins 51
Williams 71
Wilson 82, 106
Wischnitzer 58
Wolf 104
Wolff 68, 75, 77, 78
Woodard 53, 79, 170
Working distance 24
Woollam 60
Wratten: filter 26; safelight 134
Wroblewska 60
Wulff 28

Wyandt 62
Wylie 106, 109

X_2 (mitosis) 69
X-chromosome 61, 152
Xenopus 97
Xylol (Xylene) 26, 41, 48, 49, 121, 127, 128, 129, 131

Yamaha 29
Yamashita 85
Yasmineh 55
Yasuda 82
Y-chromosome 61, 81
Yeast 53, 119
Yost 71
Young 34
Yunis 55

Zakharov 64, 79
Zea 66, 69, 71, 72, 81, 85, 97, 108
Zech 63
Zelleriella 102
Zernike 30
Zinc block 90
Zirkle 38, 42, 118
Zygote 152